Embedded Digital Signal Processing Systems

EURASIP Journal on Embedded Systems

Embedded Digital Signal Processing Systems

Guest Editors: Jarmo Henrik Takala, Shuvra S. Bhattacharyya, and Gang Qu

ISBN 977-5945-84-4

Contents

Hindawi Publishing Corporation
EURASIP Journal on Embedded Systems
Volume 2007, Article ID 27517, 1 page
doi:10.1155/2007/27517

Editorial

Embedded Digital Signal Processing Systems

Jarmo Takala,[1] Shuvra S. Bhattacharyya,[2] and Gang Qu[2]

[1] *Institute of Digital and Computer Systems, Tampere University of Technology, Korkeakoulunkatu 1, 33720 Tampere, Finland*
[2] *Department of Electrical and Computer Engineering, University of Maryland, College Park, MD 20742, USA*

Received 6 March 2007; Accepted 6 March 2007

With continuing progress in VLSI and ASIC technologies, digital signal processing (DSP) algorithms have continued to find great use in increasingly wide application areas. DSP has gained popularity also in embedded systems although these systems set challenging constraints for implementations. Embedded systems contain limited resources, thus embedded DSP systems must balance tradeoffs between the requirements on computational power and computational resources. Energy efficiency has been important in battery-powered devices, but nowadays also the limited heat dissipation in small devices calls for low-power consumption. Successful implementation of DSP applications in embedded systems requires tailoring, which in turn sets challenges for design methodologies.

For this special issue, we received 14 submissions and a collection of seven papers was finally accepted. The special issue is opened by "*Observations on power-efficiency trends in mobile communication devices,*" where the authors O. Silvén and K. Jyrkkä analyze the power consumption in the current mobile communication devices. Several bottlenecks in the current implementation style have been identified, thus the paper provides a good motivation for the following papers.

In "*The Sandbridge SB3011 platform,*" the authors John Glossner et al. describe a system-on-a-chip (SoC) multiprocessor targeted as a software-defined radio platform. The platform provides solutions to the challenges in future mobile devices given in the previous paper.

"*A shared memory module for asynchronous arrays of processors*" authored by Michael J. Meeuwsen et al. considers also chip multiprocessors. The presented shared memory module can be used for interprocess communication or to increase application performance by parallelizing computation.

In "*Implementing a WLAN video terminal using UML and fully automated design flow*" by Petri Kukkala et al., an automated design flow for multiprocessor SoC is presented.

The flow is based on UML descriptions and the authors demonstrate their design flow with a design case.

Programming of chip multiprocessor platforms is considered in "*pn: a tool for improved derivation of process networks*" by Sven Verdoolaege et al. The paper discusses conversion of sequential programs to process networks allowing optimization of communication channels and buffers.

In "*A SystemC-based design methodology for digital signal processing systems,*" the authors Christian Haubelt et al. describe a design flow for SoC designs containing automatic design space exploration, performance evaluation, and automatic platform-based system generation. The design flow is based on SystemC descriptions and the presented tools can automatically detect the underlying model of computation.

Application-specific implementations are often used to speedup certain DSP tasks in embedded systems. In "*Priority-based heading one detector in H.264/AVC decoding,*" the authors Ke Xu et al. consider such implementations to speed up video decoding applications. The authors present a low-power decoder implementation for context-adaptive variable length coding defined in H.264 standard.

Jarmo Takala
Shuvra S. Bhattacharyya
Gang Qu

Hindawi Publishing Corporation
EURASIP Journal on Embedded Systems
Volume 2007, Article ID 56976, 10 pages
doi:10.1155/2007/56976

Research Article

Observations on Power-Efficiency Trends in Mobile Communication Devices

Olli Silven[1] and Kari Jyrkkä[2]

[1] Department of Electrical and Information Engineering, University of Oulu, P.O. Box 4500, 90014 Linnanmaa, Finland
[2] Technology Platforms, Nokia Corporation, Elektroniikkatie 3, 90570 Oulu, Finland

Received 3 July 2006; Revised 19 December 2006; Accepted 11 January 2007

Recommended by Jarmo Henrik Takala

Computing solutions used in mobile communications equipment are similar to those in personal and mainframe computers. The key differences between the implementations at chip level are the low leakage silicon technology and lower clock frequency used in mobile devices. The hardware and software architectures, including the operating system principles, are strikingly similar, although the mobile computing systems tend to rely more on hardware accelerators. As the performance expectations of mobile devices are increasing towards the personal computer level and beyond, power efficiency is becoming a major bottleneck. So far, the improvements of the silicon processes in mobile phones have been exploited by software designers to increase functionality and to cut development time, while usage times, and energy efficiency, have been kept at levels that satisfy the customers. Here we explain some of the observed developments and consider means of improving energy efficiency. We show that both processor and software architectures have a big impact on power consumption. Properly targeted research is needed to find the means to explicitly optimize system designs for energy efficiency, rather than maximize the nominal throughputs of the processor cores used.

1. INTRODUCTION

During the brief history of GSM mobile phones, the line widths of silicon technologies used for their implementation have decreased from 0.8 μm in the mid 1990s to around 0.13 μm in the early 21st century. In a typical phone, a basic voice call is fully executed in the baseband signal processing part, making it a very interesting reference point for comparisons as the application has not changed over the years, not even in the voice call user interface. Nokia gives the "talk-time" and "stand-by time" for its phones in the product specifications, measured according to [1] or an earlier similar convention. This enables us to track the impacts of technological changes over time.

Table 1 documents the changes in the worst case talk-times of high volume mobile phones released by Nokia between 1995 and 2003 [2], while Table 2 presents approximate characteristics of CMOS processes that have made great strides during the same period [3–5]. We make an assumption that the power consumption share of the RF power amplifier was around 50% in 1995. As the energy efficiency

of the silicon process has improved substantially from 1995 to 2003, the last phone in our table should have achieved around an 8-hour talk-time with no RF energy efficiency improvements since 1995.

During the same period (1995–2003) the gate counts of the DSP processor cores have increased significantly, but their specified power consumptions have dropped by a factor of 10 [4] from 1 mW/MIPS to 0.1 mW/MIPS. The physical sizes of the DSP cores have not essentially changed. Obviously, processor developments cannot explain why the energy efficiency of voice calls has not improved. On the microcontroller side, the energy efficiency of ARM7TMDI, for example, has improved more than 30-fold between 0.35 and 0.13 μm CMOS processes [5].

In order to offer explanations, we need to briefly analyze the underlying implementations. Figure 1 depicts streamlined block diagrams of baseband processing solutions of three product generations of GSM mobile phones. The DSP processor runs radio modem layer 1 [6] and the audio codec, whereas the microcontroller (MCU) processes layers 2 and 3 of the radio functionality and takes care of the user interface.

TABLE 1: Talk times of three mobile phones from the same manufacturer.

Year	Phone model	Talk time	Stand by time	Battery capacity
1995	2110	2 h 40 min	30 h	550 mAh
1998	6110	3 h	270 h	900 mAh
2003	6600	3 h	240 h	850 mAh

TABLE 2: Past and projected CMOS processes development.

Design rule (nm)	Supply voltage (V)	Approximate normalized power ∗ delay/gate
800 (1995)	5.0	45
500 (1998)	3.3	15
130 (2003)	1.5	1
60 (2010)	1	0.35

TABLE 3: An approximate power budget for a multimedia capable mobile phone in 384 kbit/s video streaming mode.

System component	Energy consumption (mW)
RF receiver and cellular modem	1200
Application processors and memories	600
User interface (audio, display, keyboard; with backlights)	1000
Mass memories	200
Total	3000

During voice calls, both the DSP and MCU are therefore active, while the UI introduces an almost insignificant portion of the load.

According to [7] the baseband signal processing ranks second in power consumption after RF during a voice call, and has a significant impact on energy efficiency. The baseband signal processing implementation of 1995 was based on the loop-type periodically scheduled software architecture of Figure 2 that has almost no overhead. This solution was originally dictated by the performance limitations of the processor used. Hardware accelerators were used without interrupts by relying on their deterministic latencies; this was an inherently efficient and predictable approach. On the other hand, highly skilled programmers, who understood the hardware in detail, were needed. This approach had to be abandoned after the complexity of DSP software grew due to the need to support an increasing number of features and options and the developer population became larger.

In 1998, the DSP and the microcontroller taking care of the user interface were integrated on to the same chip, and the DSP processors had become faster, eliminating some hardware accelerators [8]. Speech quality was enhanced at the cost of some additional processing on the DSP, while middleware was introduced on the microcontroller side.

The implementation of 2003 employs a preemptive operating system in the microcontroller. Basic voice call processing is still on a single DSP processor that now has a multilevel memory system. In addition to the improved voice call functionality, lots of other features are supported, including enhanced data rate for GSM evolution (EDGE), and the number of hardware accelerators increased due to higher data rates. The accelerators were synchronized with DSP tasks via interrupts. The software architecture used is ideal for large development teams, but the new functionalities, although idling during voice calls, cause some energy overhead.

The need for better software development processes has increased with the growth in the number of features in the phones. Consequently, the developers have endeavoured to preserve the active usage times of the phones at a constant level (around three hours) and turned the silicon level advances into software engineering benefits.

In the future, we expect to see advanced video capabilities and high speed data communications in mobile phones. These require more than one order of magnitude more computing power than is available in recent products, so we have to improve the energy efficiency, preferably at faster pace than silicon advances.

2. CHARACTERISTIC MODERN MOBILE COMPUTING TASKS

Mobile computing is about to enter an era of high data rate applications that require the integration of wireless wideband data modems, video cameras, net browsers, and phones into small packages with long battery powered operation times. Even the small size of phones is a design constraint as the sustained heat dissipation should be kept below 3 W [9]. In practice, much more than the capabilities of current laptop PCs is expected using around 5% of their energy and space, and at a fraction of the price. Table 3 shows a possible power budget for a multimedia phone [9]. Obviously, a 3.6 V 1000 mAh Lithium-ion battery provides only 1 hour of active operation time.

To understand how the expectations could be met, we briefly consider the characteristics of video encoding and 3GPP signal processing. These have been selected as representatives of soft and hard real time applications, and of differing hardware/software partitioning challenges.

2.1. Video encoding

The computational cost of encoding a sequence of video images into a bitstream depends on the algorithms used in the implementation and the coding standard. Table 4 illuminates the approximate costs and processing requirements of current common standards when applied to a sequence of 640-by-480 pixel (VGA) images captured at 30 frames/s. The cost of an expected "future standard" has been linearly extrapolated based on those of the past.

If a software implementation on an SISD processor is used, the operation and instructioncounts are roughly equal. This means that encoding requires the fetching and decoding

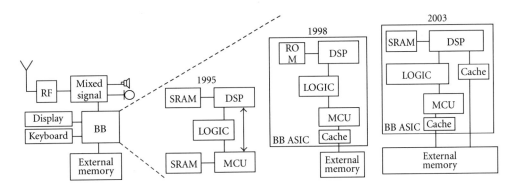

FIGURE 1: Typical implementations of mobile phones from 1995 to 2003.

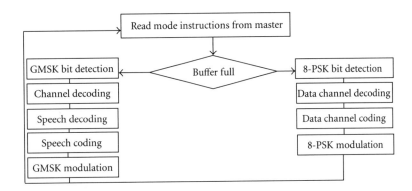

FIGURE 2: Low overhead loop-type software architecture for GSM baseband.

TABLE 4: Encoding requirements for 30 frames/s VGA video.

Video standard	Operations/pixel	Processing speed (GOPS)
MPEG-4 (1998)	200–300	2-3
H.264-AVC (2003)	600–900	6–10
"Future" (2009-10)	2000–3000	20–30

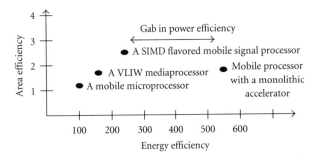

FIGURE 3: Area (Mpixels/s/mm²) and energy efficiencies (Mpixels/s/W) of comparable MPEG-4 encoder implementations.

of at least 200–300 times more instructions than pixel data. This has obvious implications from energy efficiency point of view, and can be used as a basis for comparing implementations on different programmable processor architectures.

Figure 3 illustrates the Mpixels/s per silicon area (mm²) and power (W) efficiencies of SISD, VLIW, SIMD, and the monolithic accelerator implementations of high image quality (> 34 dB PSNR) MPEG-4 VGA (advanced simple profile) video encoders. The quality requirement has been set to be relatively high so that the greediest motion estimation algorithms (such as a three-step search) are not applicable, and the search area was set to 48-by-48 pixels which fits into the on-chip RAMs of each studied processor.

All the processors are commercial and have instructions set level support for video encoding to speed-up at least summed absolute differences (SAD) calculations for 16-by-16 pixel macro blocks. The software implementation for the SISD is an original commercial one, while for VLIW and SIMD the motion estimators of commercial MPEG-4

ASP codecs were replaced by iterative full search algorithms [10, 11]. As some of the information on processors was obtained under confidentiality agreements, we are unable to name them in this paper. The monolithic hardware accelerator is a commercially available MPEG-4 VGA IP block [12] with an ARM926 core.

In the figure, the implementations have been normalized to an expected low power 1 V 60 nm CMOS process. The scaling rule assumes that power consumption is proportional to the supply voltage squared and the design rule, while the die size is proportional to the design rule squared. The original processors were implemented with 0.18 and 0.13 μm CMOS.

TABLE 5: Relative instruction fetch rates and control unit sizes versus area and energy efficiencies.

Solution	Instruction fetch/decode rate	Control unit size	Area efficiency	Energy efficiency
SISD	Operation rate	Relatively small	Lowest	Lowest
VLIW	Operation rate	Relatively small	Average	Average
SIMD	Less than operation rate	Relatively small	Highest	Good
Monolithic accelerator	Very low (control code)	Very small	Average	Highest

We notice a substantial gap in energy efficiency between the monolithic accelerator and the programmed approaches. For instance, around 40 mW of power is needed for encoding 10 Mpixels/s using the SIMD extended processor, while the monolithic accelerator requires only 16 mW. In reality, the efficiency gap is even larger as the data points have been determined using only a single task on each processor. In practice, the processors switch contexts between tasks and serve hardware interrupts, reducing the hit rates of instruction and data caches, and the branch prediction mechanism. This may easily drop the actual processing throughput by half, and, respectively, lowers the energy efficiency.

The sizes of the control units and instruction fetch rates needed for video encoding appear to explain the data points of the programmed solutions as indicated by Table 5. The SISD and VLIW have the highest fetch rates, while the SIMD has the lowest one, contributing to energy efficiency. The execution units of the SIMD and VLIW occupy relatively larger portions of the processor chips: this improves the silicon area efficiency as the control part is overhead. The monolithic accelerator is controlled via a finite state machine, and needs processor services only once every frame, allowing the processor to sleep during frames.

In this comparison, the silicon area efficiency of the hardware accelerated solution appears to be reasonably good, as around 5 mm^2 of silicon is needed for achieving real-time encoding for VGA sequences. This is better than for the SISD (9 mm^2) and close to the SIMD (around 4 mm^2). However, the accelerator supports only one video standard, while support for another one requires another accelerator, making hardware acceleration in this case the most inefficient approach in terms of silicon area and reproduction costs.

Consequently, it is worth considering whether the video accelerator could be partitioned in a manner that would enable re-using its components in multiple coding standards. The speed-up achieved from these finer grained approaches needs to be weighted against the added overheads such as the typical 300 clock cycle interrupt latency that can become significant if, for example, an interrupt is generated for each 16-by-16 pixel macroblock of the VGA sequence.

An interesting point for further comparisons is the hibrid-SOC [13], that is, the creation of one research team. It is a multicore architecture, based on three programmable dedicated core processors (SIMD, VLIW, and SISD), intended for video encoding and decoding, and other high bandwidth applications. Based on the performance and implementation data, it comes very close to the VLIW device in Figure 2 when scaled to the 60 nm CMOS technology of Table 2, and it could rank better if explicitly designed for low power operation.

TABLE 6: 3GPP receiver requirements for different channel types.

Channel type	Data rate	Processing speed (GOPS)
Release 99 DCH channel	0.384 Mbps	1-2
Release 5 HSDPA channel	14.4 Mbps	35–40
"Future 3.9G" OFDM channel	100 Mbps	210–290

2.2. 3GPP baseband signal processing

Based on its timing requirements, the 3GPP baseband signal processing chain is an archetypal hard real-time application that is further complicated by the heavy computational requirements shown in Table 6 for the receiver. The values in the table have been determined for a solution using turbo decoding and they do not include chip-level decoding and symbol level combining that further increase the processing needs.

The requirements of the high speed downlink packet access (HSDPA) channel that is expected to be introduced in mobile devices in the near future characterize current acute implementation challenges. Interestingly, the operation counts per received bit for each channel are roughly in the same magnitude range as with video encoding.

Figure 4 shows the organization of the 3GPP receiver processing and illuminates the implementation issues. The receiver data chain has time critical feedback loops implemented in the software; for instance, the control channel HS-SCCH is used to control what is received, and when, on the HS-DSCH data channel. Another example is the power control information decoded from "release 99 DSCH" channel that is used to regulate the transmitter power 1500 times per second. Furthermore, the channel code rates, channel codes, and interleaving schemes may change anytime, requiring software control for reconfiguring the hardware blocks of the receiver, although for clarity this is not indicated in the diagram.

The computing power needs of 3GPP signal processing have so far been satisfied only by hardware at an acceptable

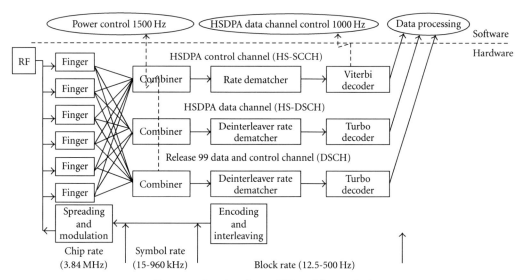

FIGURE 4: Receiver for a 3GPP mobile terminal.

energy efficiency level. Software implementations for turbo decoding that meet the speed requirement do exist; for instance, in [14] the performance of analog devices' Tiger-SHARC DSP processor is demonstrated. However, it falls short of the energy efficiency needed in phones and is more suitable for base station use.

For energy efficiency, battery powered systems have to rely on hardware, while the tight timings demand the employment of fine grained accelerators. A resulting large interrupt load on the control processors is an undesired side effect. Coarser grain hardware accelerators could reduce this overhead, but this is an inflexible approach and riskier when the channel specifications have not been completely frozen, but the development of hardware must begin.

With reservations on the hard real-time features, the results of the above comparison on the relative efficiencies of processor architectures for video encoding can be extended to 3GPP receivers. Both tasks have high processing requirements and the grain size of the algorithms is not very different, so they could benefit from similar solutions that improve hardware reuse and energy efficiency. In principle, the processor resources can be used more efficiently with the softer real-time demands of video coding, but if fine grained acceleration is used instead of a monolithic solution, it becomes a hard real-time task.

3. ANALYSIS OF THE OBSERVED DEVELOPMENT

Based on our understanding, there is no single action that could improve the talk-times of mobile phones and usage times of future applications. Rather there are multiple interacting issues for which balanced solutions must be found. In the following, we analyze some of the factors considered to be essential.

3.1. Changes in voice call application

The voice codec in 1995 required around 50% of the operation count of the more recent codec that provides improved voice quality. As a result, the computational cost of the basic GSM voice call may have even more than doubled [15]. This qualitative improvement has in part diluted the benefits obtained through advances in semiconductor processes, and is reflected by the talk-time data given for the different voice codec by mobile terminal manufacturers. It is likely that the computational costs of voice calls will increase even in the future with advanced features.

3.2. The effect of preemptive real-time operating systems

The dominating scheduling principle used in embedded systems is "rate monotonic analysis (RMA)" that assigns higher static priorities for tasks that execute at higher rates. When the number of tasks is large, utilizing the processor at most up to 69% guarantees that all deadlines are met [16]. If more processor resources are needed, then more advanced analysis is needed to learn whether the scheduling meets the requirements.

In practice, both our video and 3GPP baseband examples are affected by this law. A video encoder, even when fully implemented in software, is seldom the only task in the processor, but shares its resources with a number of other tasks. The 3GPP baseband processing chain consists of several simultaneous tasks due to time critical hardware/software interactions.

With RMA, the processor utilization limit alone may demand even 40% higher clock rates than was necessary with the static cyclic scheduling used in early GSM phones in which the clock could be controlled very flexibly. Now, due to the scheduling overhead that has to be added to the task durations, a 50% clock frequency increase is close to reality.

We admit that this kind of comparison is not completely fair. Static cyclic scheduling is no longer usable as it is unsuitable for providing responses for sporadic events within a short fixed time, as required by the newer features of the

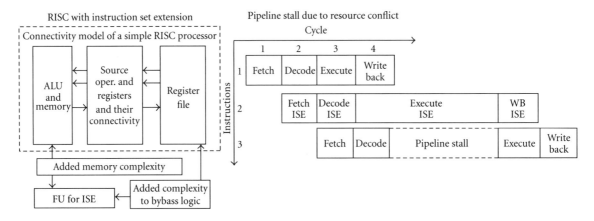

FIGURE 5: Hardware acceleration via instruction set extension.

phones. The use of dynamic priorities and earliest-deadline-first (EDF) or least-slack algorithm [17] would improve processor utilization over RMA, although this would be at the cost of slightly higher scheduling overheads that can be significant if the number of tasks is large. Furthermore, embedded software designers wish to avoid EDF scheduling, because variations in cache hit ratios complicate the estimation of the proximity of deadlines.

3.3. The effect of context switches on cache and processor performance

The instruction and data caches of modern processors improve energy efficiency when they perform as intended. However, when the number of tasks and the frequency of context switches is high, the cache-hit rates may suffer. Experiments [18] carried out using the MiBench [19] embedded benchmark suite on an MIPS 4KE-type instruction set architecture revealed that with a 16 kB 4-way set associative instruction cache the hit-rate averaged around 78% immediately after context switches and 90% after 1000 instructions, while 96% was reached after the execution of 10 000 instructions.

Depending on the access time differential between the main memory and the cache, the performance impact can be significant. If the processor operates at 150 MHz with a 50-nanosecond main memory and an 86% cache hit rate, the execution time of a short task slice (say 2000 instructions) almost doubles. Worst of all, the execution time of the same piece of code may fluctuate from activation to activation, causing scheduling and throughput complications, and may ultimately force the system implementers to increase the processor clock rate to ensure that the deadlines are met.

Depending on the implementations, both video encoder and 3GPP baseband applications operate in an environment that executes up to tens of thousands of interrupts and context switches in a second. Although this facilitates the development of systems with large teams, the approach may have a significant negative impact on energy efficiency.

More than a decade ago (1991), Mogul and Borg [20] made empirical measurements on the effects of context switches on cache and system performance. After a partial reproduction of their experiments on a modern processor, Sebek [21] comments "it is interesting that the cache related preemption delay is almost the same," although the processors have became a magnitude faster. We may make a similar observation about GSM phones and voice calls: current implementations of the same application require more resources than in the past. This cycle needs to be broken in future mobile terminals and their applications.

3.4. The effect of hardware/software interfacing

The designers of mobile phones aim to create common platforms for product families. They define application programming interfaces that remain the same, regardless of system enhancements and changes in hardware/software partitioning [8]. This has made middleware solutions attractive, despite worries over the impact on performance. However, the low level hardware accelerator/software interface is often the most critical one.

Two approaches are available for interfacing hardware accelerators to software. First, a hardware accelerator can be integrated into the system as an extension to the instruction set, as illustrated with Figure 5. In order to make sense, the latency of the extension should be in the same range as the standard instructions, or, at most, within a few instruction cycles, otherwise the interrupt response time may suffer. Short latency often implies large gate count and high bus bandwidth needs that reduce the economic viability of the approach, making it a rare choice in mobile phones.

Second, an accelerator may be used in a peripheral device that generates an interrupt after completing its task. This principle is demonstrated in Figure 6, which also shows the role of middleware in hiding details of the hardware. Note that the legend in the picture is in the order of priority levels.

If the code in the middleware is not integrated into the task, calls to middleware functions are likely to reduce the cache hit rate. Furthermore, to avoid high interrupt overheads, the execution time of the accelerators should

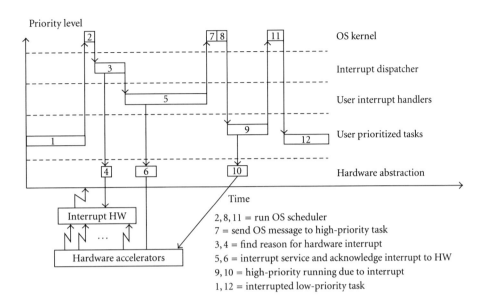

FIGURE 6: Controlling an accelerator interfaced as a peripheral device.

TABLE 7: Energy efficiencies and silicon areas of ARM processors.

Processor	Processor max. clock frequency (MHz)	Silicon area (mm²)	Power consumption (mW/MHz)
ARM9 (926EJ-S)	266	4.5	0.45
ARM10 (1022E)	325	6.9	0.6
ARM11 (1136J-S)	550	5.55	0.8

preferably be thousands of clock cycles. In practice, this approach is used even with rather short latency accelerators, as long as it helps in achieving the total performance target. The latencies from middleware, context switches, and interrupts have obvious consequences for energy efficiency.

Against this background, it is logical that the monolithic accelerator turned out to be the most energy efficient solution for video encoding in Figure 3. From the point of view, the 3GPP baseband a key to energy efficient implementation in a given hardware lies in pushing down the latency overheads.

It is rather interesting that anything in between 1-2 cycle instruction set extensions and peripheral devices executing thousands of cycles can result in grossly inefficient software. If the interrupt latency in the operating system environment is around 300 cycles and 50 000 interrupts are generated per second, 10% of the 150 MHz processor resources are swallowed by this overhead alone, and on top of this we have middleware costs. Clearly, we have insufficient expertise in this bottleneck area that falls between hardware and software, architectures and mechanisms, and systems and components.

3.5. The effect of processor hardware core solutions

Current DSP processor execution units are deeply pipelined to increase instruction execution rates. In many cases, however, DSP processors are used as control processors and have to handle large interrupt and context switch loads. The result is a double penalty: the utilization of the pipeline decreases and the control code is inefficient due to the long pipeline. For instance, if a processor has a 10-level pipeline and 1/50 of the instructions are unconditional branches, almost 20% of the cycles are lost. Improvements offered by the branch prediction capabilities are diluted by the interrupts and context switches.

The relative sizes of control units of typical low power DSP processors and microcontrollers have increased during recent years due to deeper pipelining. However, when executing control code, most of the processor is unused. This situation is encountered with all fine grained hardware accelerator-based implementations regardless of whether they are video encoder or 3GPP baseband solutions. Obviously, rethinking the architectures and their roles in the system implementations is necessary. To illustrate the impact of increasing processor complexity on the energy efficiency, Table 7 shows the characteristics of 32-bit ARM processors implemented using a 130 nm CMOS process [5]. It is apparent that the energy efficiencies of processor designs are increasing, but this development has been masked by silicon process developments. Over the past ten years the relative efficiency appears to have slipped approximately by a factor of two.

TABLE 8: Approximate efficiency degradations.

Degradation cause	Low estimate	Probable degradation
Computational cost of voice call application	2	2.5
Operating system and interrupt overheads	1.4	1.6
API and middleware costs	1.2	1.5
Execution time jitter provisioning	1.3	2
Processor energy/instruction	1.8	2.5
Execution pipeline overheads	1.2	1.5
Total (multiplied)	9.4	45

3.6. Summary of relative performance degradations

When the components of the above analysis are combined as shown in Table 8, they result in a degradation factor of at least around 9-10, but probably around 45. These are relative energy efficiency degradations and illustrate the traded-off energy efficiency gains at the processing system level. The probable numbers appear to be in line with the actual observed development.

It is acknowledged in industry that approaches in system development have been dictated by the needs of software development that has been carried out using the tools and methods available. Currently, the computing needs are increasing rapidly, so a shift of focus to energy efficiency is required. Based on Figure 3, using suitable programmable processor architectures can improve the energy efficiency significantly. However, in baseband signal processing the architectures used already appear fairly optimal. Consequently, other means need to be explored too.

4. DIRECTIONS FOR RESEARCH AND DEVELOPMENT

Looking back to the phone of 1995 in Table 1, we may consider what should have been done to improve energy efficiency at the same rate as silicon process improvement. Obviously, due to the choices made by system developers, most of the factors that degrade the relative energy efficiency are software related. However, we do not demand changes in software development processes or architectures that are intended to facilitate human effort. So solutions should primarily be sought from the software/hardware interfacing domain, including compilation, and hardware solutions that enable the building of energy efficient software systems.

To reiterate, the early baseband software was effectively multi-threaded, and even simultaneously multithreaded with hardware accelerators executing parallel threads, without interrupt overhead, as shown in Figure 7. In principle, a suitable compiler could have replaced manual coding in creating the threads, as the hardware accelerators had deterministic latencies. However, interrupts were introduced and later solutions employed additional means to hide the hardware from the programmers.

Having witnessed the past choices, their motivations, and outcomes, we need to ask whether compilers could be used to hide hardware details instead of using APIs and middleware. This approach could in many cases cut down the number of interrupts, reduce the number of tasks and context switches, and improve code locality— all improving processor utilization and energy efficiency. Most importantly, hardware accelerator aware compilation would bridge the software efficiency gap between instruction set extensions and peripheral devices, making "medium latency" accelerators attractive. This would help in cutting the instruction fetch and decoding overheads.

The downside of a hardware aware compilation approach is that the binary software may no longer be portable, but this is not important for the baseband part. A bigger issue is the paradigm change that the proposed approach represents. Compilers have so far been developed for processor cores; now they would be needed for complete embedded systems. Whenever the platform changes, the compiler needs to be upgraded, while currently the changes are concentrated on the hardware abstraction functionality.

Hardware support for simultaneous fine grained multithreading is an obvious processor core feature that could contribute to energy efficiency. This would help in reducing the costs of scheduling.

Another option that could improve energy efficiency is the employing of several small processor cores for controlling hardware accelerators, rather that a single powerful one. This simplifies real-time system design and reduces the total penalty from interrupts, context switches, and execution time jitter. To give a justification for this approach, we again observe that the W/MHz figures for the 16-bit ARM7/TDMI dropped by factor 35 between 0.35 and 0.13 μm CMOS processes [5]. Advanced static scheduling and allocation techniques [22] enable constructing efficient tools for this approach, making it very attractive.

5. SUMMARY

The energy efficiency of mobile phones has not improved at the rate that might have been expected from the advances in silicon processes, but it is obviously at a level that satisfies most users. However, higher data rates and multimedia applications require significant improvements, and encourage us to reconsider the ways software is designed, run, and interfaced with hardware.

Significantly improved energy efficiency might be possible even without any changes to hardware by using software solutions that reduce overheads and improve processor utilization. Large savings can be expected from applying architectural approaches that reduce the volume of instructions fetched and decoded. Obviously, compiler technology is the key enabler for improvements.

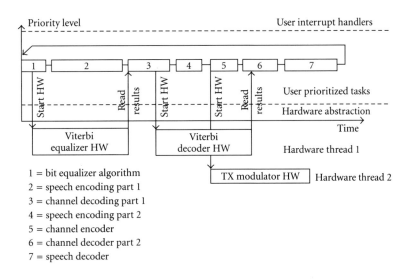

FIGURE 7: The execution threads of an early GSM mobile phone.

ACKNOWLEDGMENTS

Numerous people have directly and indirectly contributed to this paper. In particular, we wish to thank Dr. Lauri Pirttiaho for his observations, comments, questions, and expertise, and Professor Yrjö Neuvo for advice, encouragement, and long-time support, both from the Nokia Corporation.

REFERENCES

[1] GSM Association, "TW.09 Battery Life Measurement Technique," 1998, http://www.gsmworld.com/documents/index.shtml.

[2] Nokia, "Phone models," http://www.nokia.com/.

[3] M. Anis, M. Allam, and M. Elmasry, "Impact of technology scaling on CMOS logic styles," *IEEE Transactions on Circuits and Systems II: Analog and Digital Signal Processing*, vol. 49, no. 8, pp. 577–588, 2002.

[4] G. Frantz, "Digital signal processor trends," *IEEE Micro*, vol. 20, no. 6, pp. 52–59, 2000.

[5] The ARM foundry program, 2004 and 2006, http://www.arm.com/.

[6] 3GPP: TS 05.01, "Physical Layer on the Radio Path (General Description)," http://www.3gpp.org/ftp/Specs/html-info/0501.htm.

[7] J. Doyle and B. Broach, "Small gains in power efficiency now, bigger gains tomorrow," *EE Times*, 2002.

[8] K. Jyrkkä, O. Silven, O. Ali-Yrkkö, R. Heidari, and H. Berg, "Component-based development of DSP software for mobile communication terminals," *Microprocessors and Microsystems*, vol. 26, no. 9-10, pp. 463–474, 2002.

[9] Y. Neuvo, "Cellular phones as embedded systems," in *Proceedings of IEEE International Solid-State Circuits Conference (ISSCC '04)*, vol. 1, pp. 32–37, San Francisco, Calif, USA, February 2004.

[10] X. Q. Gao, C. J. Duanmu, and C. R. Zou, "A multilevel successive elimination algorithm for block matching motion estimation," *IEEE Transactions on Image Processing*, vol. 9, no. 3, pp. 501–504, 2000.

[11] H.-S. Wang and R. M. Mersereau, "Fast algorithms for the estimation of motion vectors," *IEEE Transactions on Image Processing*, vol. 8, no. 3, pp. 435–438, 1999.

[12] 5250 VGA encoder, 2004, http://www.hantro.com/en/products/codecs/hardware/5250.html.

[13] S. Moch, M. Bereković, H. J. Stolberg, et al., "HIBRID-SOC: a multi-core architecture for image and video applications," *ACM SIGARCH Computer Architecture News*, vol. 32, no. 3, pp. 55–61, 2004.

[14] K. K. Loo, T. Alukaidey, and S. A. Jimaa, "High performance parallelised 3GPP turbo decoder," in *Proceedings of the 5th European Personal Mobile Communications Conference (EPMCC '03)*, Conf. Publ. no. 492, pp. 337–342, Glasgow, UK, April 2003.

[15] R. Salami, C. Laflamme, B. Bessette, et al., "Description of GSM enhanced full rate speech codec," in *Proceedings of the IEEE International Conference on Communications (ICC '97)*, vol. 2, pp. 725–729, Montreal, Canada, June 1997.

[16] M. H. Klein, *A Practitioner's Handbook for Real-Time Analysis*, Kluwer, Boston, Mass, USA, 1993.

[17] M. Spuri and G. C. Buttazzo, "Efficient aperiodic service under earliest deadline scheduling," in *Proceedings of Real-Time Systems Symposium*, pp. 2–11, San Juan, Puerto Rico, USA, December 1994.

[18] J. Stärner and L. Asplund, "Measuring the cache interference cost in preemptive real-time systems," in *Proceedings of the ACM SIGPLAN Conference on Languages, Compilers, and Tools for Embedded Systems (LCTES '04)*, pp. 146–154, Washington, DC, USA, June 2004.

[19] M. R. Gathaus, J. S. Ringenberg, D. Ernst, T. M. Austen, T. Mudge, and R. B. Brown, "MiBench: a free, commercially representative embedded benchmark suite," in *Proceedings of the 4th Annual IEEE International Workshop on Workload Characterization (WWC-4 '01)*, pp. 3–14, Austin, Tex, USA, December 2001.

[20] J. C. Mogul and A. Borg, "The effect of context switches on cache performance," in *Proceedings of the 4th International Conference on Architectural Support for Programming Languages and Operating Systems (ASPLOS '91)*, pp. 75–84, Santa Clara, Calif, USA, April 1991.

[21] F. Sebek, "Instruction cache memory issues in real-time systems," Technology Licentiate thesis, Department of Computer Science and Engineering, Mälardalen University, Västerås, Sweden, 2002.

[22] S. Sriram and S. S. Bhattacharyya, *Embedded Multiprocessors: Scheduling and Synchronization*, Marcel Dekker, New York, NY, USA, 2000.

Hindawi Publishing Corporation
EURASIP Journal on Embedded Systems
Volume 2007, Article ID 56467, 16 pages
doi:10.1155/2007/56467

Research Article

The Sandbridge SB3011 Platform

John Glossner, Daniel Iancu, Mayan Moudgill, Gary Nacer, Sanjay Jinturkar, Stuart Stanley, and Michael Schulte

Sandbridge Technologies, Inc., 1 North Lexington Avenue, White Plains, NY 10601, USA

Received 1 August 2006; Revised 18 January 2007; Accepted 20 February 2007

Recommended by Jarmo Henrik Takala

This paper describes the Sandbridge Sandblaster real-time software-defined radio platform. Specifically, we describe the SB3011 system-on-a-chip multiprocessor. We describe the software development system that enables real-time execution of communications and multimedia applications. We provide results for a number of interesting communications and multimedia systems including UMTS, DVB-H, WiMAX, WiFi, and NTSC video decoding. Each processor core achieves 600 MHz at 0.9 V operation while typically dissipating 75 mW in 90 nm technology. The entire chip typically dissipates less than 500 mW at 0.9 V.

1. INTRODUCTION

Performance requirements for mobile wireless communication devices have expanded dramatically since their inception as mobile telephones. Recent carrier offerings with multiple communication systems and handover from cellular to broadband suggest that some consumers are requesting convergence devices with full data and voice integration. The proliferation of cameras and Internet access in cell phones also suggests a variety of computationally intense features and applications such as web browsing, MP3 audio, and MPEG4 video are needed. Moreover, consumers want these wireless subscriber services to be accessible at all times anywhere in the world. Such complex functionality requires high computing capability at low power consumption; adding new features requires adding computing capacity.

The technologies necessary to realize true broadband wireless handsets and systems presenting unique design challenges if extremely power efficient, yet high-performance, broadband wireless terminals are to be realized. The design tradeoffs and implementation options inherent in meeting such demands highlight the extremely onerous requirements for next generation baseband processors. Tremendous hardware and software challenges exist to realize convergence devices.

The increasing complexities of mobile terminals and a desire to generate multiple versions with increasing features for handsets have led to the consideration of a software-defined radio- (SDR-) based approach in the wireless industry. The previous generation of mobile terminals was primarily designed for use in geographically restricted areas where growth of the wireless industry was dependant upon signing up new users. The penetration levels in European and Asian countries are high and new revenue streams (from technologies such as 3G) have been slow to materialize for a variety of complex reasons. True convergence of multimedia, cellular, location and connectivity technologies is expensive, time consuming, and complex at all levels of development—not only mobile terminals, but infrastructure as well. Moreover, the standards themselves have failed to converge, which has led to multiple market segments. In order to maintain market share, a handset development company must use multiple platforms each of which may be geographically specific supporting multiple combinations of communications systems. This requires some handset companies to support multiple platforms and multiple hardware solutions from multiple technology suppliers.

1.1. SDR-based approach

Building large parallel processing systems is a difficult task. Programming them efficiently is even more challenging. When nonassociative digital signal processing (DSP) arithmetic is included, the challenge of automated software development for a complex chip multiprocessor (CMP) system is amplified.

Early software-defined radio (SDR) platforms were often built out of discrete processors and FPGAs that were integrated on a card. More recently a trend has been to integrate multiple processors on a single chip creating SDR CMP systems. The SDR Forum [1] defines five tiers of solutions. Tier-0 is a traditional radio implementation in hardware. Tier-1, software-controlled radio (SCR), implements the control features for multiple hardware elements in software. Tier-2, software-defined radio (SDR), implements modulation and baseband processing in software but allows for multiple frequency fixed function RF hardware. Tier-3, ideal software radio (ISR), extends programmability through the RF with analog conversion at the antenna. Tier-4, ultimate software radio (USR), provides for fast (millisecond) transitions between communications protocols in addition to digital processing capability.

The advantages of reconfigurable SDR solutions versus hardware solutions are significant. First, reconfigurable solutions are more flexible allowing multiple communication protocols to dynamically execute on the same transistors thereby reducing hardware costs. Specific functions such as filters, modulation schemes, encoders/decoders can be reconfigured adaptively at run time. Second, several communication protocols can be efficiently stored in memory and coexist or execute concurrently. This significantly reduces the cost of the system for both the end user and the service provider. Third, remote reconfiguration provides simple and inexpensive maintenance and feature upgrades. This also allows service providers to differentiate products after the product is deployed. Fourth, the development time of new and existing communications protocols is significantly reduced providing an accelerated time to market. Development cycles are not limited by long and laborious hardware design cycles. With SDR, new protocols are quickly added as soon as the software is available for deployment. Fifth, SDR provides an attractive method of dealing with new standards releases while assuring backward compatibility with existing standards.

SDR enabling technologies also have significant advantages from the consumer perspective. First, mobile terminal independence with the ability to "choose" desired feature sets is provided. As an example, the same terminal may be capable of supporting a superset of features but the consumer only pays for features that they are interested in using. Second, global connectivity with the ability to roam across operators using different communications protocols can be provided. Third, future scalability and upgradeability provide for longer handset lifetimes.

1.2. Processor background

In this section we define a number of terms and provide background information on general purpose processors, digital signal processors, and some of the workload differences between general purpose computers and real-time embedded systems.

The *architecture* of a computer system is the minimal set of properties that determine what programs will run and what results they will produce [2]. It is the contract between the programmer and the hardware. Every computer is an interpreter of its *machine language*—that representation of programs that resides in memory and is interpreted (executed) directly by the (host) hardware.

The logical organization of a computer's dataflow and controls is called the *implementation or microarchitecture*. The physical structure embodying the implementation is called the *realization*. The architecture describes what happens while the implementation describes how it is made to happen. Programs of the same architecture should run unchanged on different implementations. An architectural function is *transparent* if its implementation does not produce any architecturally visible side effects. An example of a nontransparent function is the load delay slot made visible due to pipeline effects. Generally, it is desirable to have transparent implementations. Most DSP and VLIW implementations are not transparent and therefore the implementation affects the architecture [3].

Execution predictability in DSP systems often precludes the use of many general-purpose design techniques (e.g., speculation, branch prediction, data caches, etc.). Instead, classical DSP architectures have developed a unique set of performance-enhancing techniques that are optimized for their intended market. These techniques are characterized by hardware that supports efficient filtering, such as the ability to sustain three memory accesses per cycle (one instruction, one coefficient, and one data access). Sophisticated addressing modes such as bit-reversed and modulo addressing may also be provided. Multiple address units operate in parallel with the datapath to sustain the execution of the inner kernel.

In classical DSP architectures, the execution pipelines were visible to the programmer and necessarily shallow to allow assembly language optimization. This programming restriction encumbered implementations with tight timing constraints for both arithmetic execution and memory access. The key characteristic that separates modern DSP architectures from classical DSP architectures is the focus on compilability. Once the decision was made to focus the DSP design on programmer productivity, other constraining decisions could be relaxed. As a result, significantly longer pipelines with multiple cycles to access memory and multiple cycles to compute arithmetic operations could be utilized. This has yielded higher clock frequencies and higher performance DSPs.

In an attempt to exploit instruction level parallelism inherent in DSP applications, modern DSPs tend to use VLIW-like execution packets. This is partly driven by real-time requirements which require the worst-case execution time to be minimized. This is in contrast with general purpose CPUs which tend to minimize average execution times. With long pipelines and multiple instruction issue, the difficulties of attempting assembly language programming become apparent. Controlling dependencies between upwards of 100 inflight instructions is not an easy task for a programmer. This is exactly the area where a compiler excels.

One challenge of using some VLIW processors is large program executables (code bloat) that result from independently specifying every operation with a single instruction. As an example, a VLIW processor with a 32-bit basic instruction width may require 4 instructions, 128 bits, to specify 4 operations. A vector encoding may compute many more operations in as few as 21 bits (e.g., multiply two 4-element vectors, saturate, accumulate, and saturate).

Another challenge of some VLIW implementations is that they may have excessive register file write ports. Because each instruction may specify a unique destination address and all the instructions are independent, a separate port may be provided for the target of each instruction. This can result in high power dissipation, which is unacceptable for handset applications.

To help overcome problems with code bloat and excessive write ports, recent VLIW DSP architectures, such as OnDSP [4], the embedded vector processor (EVP) [5], and the synchronous transfer architecture (STA) [6], provide vector operations, specialized instructions for multimedia and wireless communications, and multiple register files.

A challenge of visible pipeline machines (e.g., most DSPs and VLIW processors) is interrupt response latency. It is desirable for computational datapaths to remain fully utilized. Loading new data while simultaneously operating on current data is required to maintain execution throughput. However, visible memory pipeline effects in these highly parallel inner loops (e.g., a load instruction followed by another load instruction) are not typically interruptible because the processor state cannot be restored. This requires programmers to break apart loops so that worst-case timings and maximum system latencies may be acceptable.

Signal processing applications often require both computations and control processing. Control processing is often amenable to RISC-style architectures and is typically compiled directly from C code. Signal processing computations are characterized by multiply-accumulate intensive functions executed on fixed point vectors of moderate length. Therefore, a DSP requires support for such fixed point saturating computations. This has traditionally been implemented using one or more multiply accumulate (MAC) units. In addition, as the saturating arithmetic is nonassociative, parallel operations on multiple data elements may result in different results from serial execution. This creates a challenge for high-level language implementations that specify integer modulo arithmetic. Therefore, most DSPs have been programmed using assembly language.

Multimedia adds additional requirements. A processor which executes general purpose programs, signal processing programs, and multimedia programs (which may also be considered to be signal processing programs) is termed a convergence processor. Video, in particular, requires high performance to allow the display of movies in real-time. An additional trend for multimedia applications is Java execution. Java provides a user-friendly interface, support for productivity tools and games on the convergence device.

The problems associated with previous approaches require a new architecture to facilitate efficient convergence applications processing. Sandbridge Technologies has developed a new approach that reduces both hardware and software design challenges inherent in real-time applications like

SDR and processing of streaming data in convergence services.

In the subsequent sections, we describe the Sandbridge SB3011 low-power platform, the architecture and implementation, the programming tools including an automatically multithreading compiler, and SDR results.

2. THE SB3011 SDR PLATFORM

Motivated by the convergence of communications and multimedia processing, the Sandbridge SB3011 was designed for efficient software execution of physical layer, protocol stacks, and multimedia applications. The Sandbridge SDR platform is a Tier-2 implementation as defined by the SDR Forum. Figure 1 shows the SB3011 implementation. It is intended for handset markets. The main processing complex includes four DSPs [7] each running at a minimum of 600 MHz at 0.9 V. The chip is fabricated in 90 nm technology. Each DSP is capable of issuing multiple operations per cycle including data parallel vector operations. The microarchitecture of each DSP is 8-way multithreaded allowing the SB3011 to simultaneously execute up to 32 independent instruction streams each of which may issue vector operations.

2.1. DSP complex

Each DSP has a level-1 (L1) 32 KB set-associative instruction cache and an independent L1 64 KB data memory which is not cached. In addition a noncached global level-2 (L2) 1 MB memory is shared among all processors. The implementation guarantees no pipeline stalls for L1 memory accesses (see Section 4). For external memory accesses or L2 accesses only the thread that issued the load request stalls. All other threads continue executing independent of which processor issued the memory request.

The Sandblaster DSP is a true architecture in the sense that from the programmer's perspective each instruction completes prior to the next instruction issuing—on a per thread basis.

The processors are interconnected through a deterministic and opportunistic unidirectional ring network. The interconnection network typically runs at half the processor speed. The ring is time-division multiplexed and each processor may request a slot based on a proprietary algorithm. Communications between processors is primarily through shared memory.

The processor's instruction set provides synchronization primitives such as load locked and store conditional. Since all data memory is noncached, there are no coherence issues.

2.2. ARM and ARM peripherals

In addition to the parallel multithreaded DSP complex, there is an entire ARM complex with all the peripherals necessary to implement input/output (I/O) devices in a smart phone. The processor is an ARM926EJ-S running at up to 300 MHz. The ARM has 32 kB instruction and 32 kB data cache memories. There is an additional 64 kB of scratch

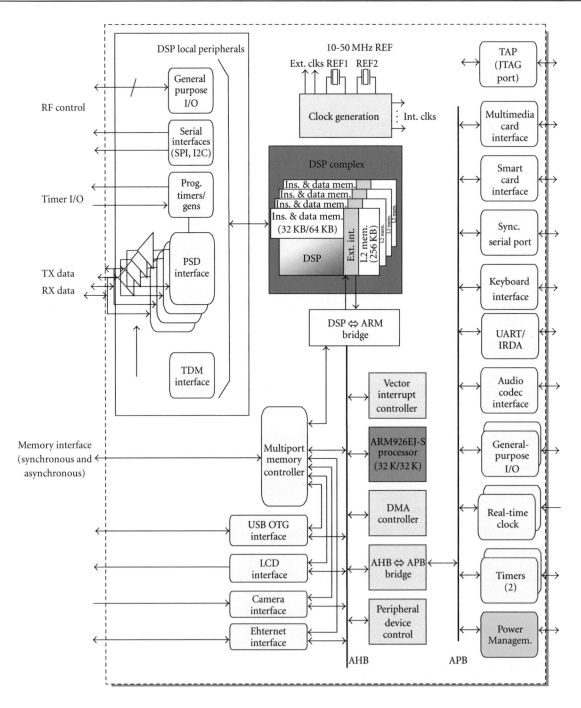

FIGURE 1: Sandblaster SB3011 chip.

memory partitioned as 32 kB instruction and 32 kB data. Sandbridge has ported Linux to this platform and the processor functions as the user interface (UI or sometimes MMI) for smart phone applications.

Using an AMBA advanced high-speed bus (AHB) and advanced peripheral bus (APB), the system is able to support the processing of multiple concurrent data interfaces. Attached to the APB is a multimedia card (MMC) and secure digital card (SD Card) interface for connecting external Flash storage. Keyboard and mouse interfaces are included along

with multiple UARTs and an IRDA interface. Audio and microphone PCM data is supported through an AC-97 interface which connects directly to an external codec. A number of other general peripherals are also supported on the APB including a real-time clock and general purpose timers, which are used to keep system time.

The AHB is used to move high-speed data such as memory requests and video into the chip for processing or out of the chip for display or storage. A direct connection to an LCD display is provided and can be accessed from either the ARM or DSP processors. Similarly, a high-speed camera interface is provided to capture raw video data that may be encoded or processed in the DSP or ARM processors.

The SB3011 includes a full USB 2.0 on-the-go (OTG) implementation for seamless connection to printers, cameras, storage, or other USB devices. An Ethernet interface is also included on the chip for wired local area network (LAN) connections.

2.3. External memory

External memory requests which both the ARM and DSPs can initiate are routed through a multiport memory controller attached to the AHB. The external memory can be synchronous or asynchronous. Typical memories include Flash, SDRAM, DRAM, and SRAM. The controller supports multiple simultaneous requests whether generated through direct memory access (DMA) devices (both for the ARM and DSP) or by a direct address from the processors. External memory requests are managed by an arbitration controller which ensures priority and fairness for the bus transactions. All external memory is mapped into a 32-bit global address space and is shared among all processors.

Device processors are booted from external memory in a two-step sequence: ARM followed by the DSPs. Once the device is released from reset, the ARM processor begins execution from one of the memory controller's memory ports (typically the port connected to Flash memory on the card). The ARM then executes a device initialization routine, which configures the necessary internal device peripherals and the execution vectors for the DSPs. Once complete, the DSPs are enabled and each processor begins executing the Sandbridge Operating System (SBOS), which may be in Flash or other memory.

2.4. DSP peripherals

In addition to the ARM peripherals, there are a number of DSP specific peripherals primarily intended for moving data to and from external radio frequency (RF) devices, time division multiplexed (TDM) voice and data devices (e.g., T1/E1), and other peripherals. These peripherals interface directly to the DSPs' L2 memory subsystem through the multiple parallel streaming data (PSD) or TDM interfaces. Four half-duplex PSD interfaces are provided, each supporting up to 16-bit data samples. PSD data is latched or transmitted by the device on both edges of its respective clock, thus realizing two data streams per interface (typically I and Q streams to and from an RF's analog-front-end device). Four serial TDM interfaces are provided, each of which capable of up to 1024 channels, for an aggregate 32 k samples per second throughput. Support for synchronization of transmitted or received data bursts is accomplished through the use of dedicated I/O timers. When configured, these timers can be operated with an external (system) clock source and are internally used to gate the DMA transfers on the PSD interfaces. This feature is important for slot-based communications systems such as GSM.

A number of other interfaces are provided for general purpose control of external components typically found in smart phones. These include general purpose timers which can be used as external clocks, SPI, and I2C buses which are common in RF control logic, and general purpose I/O. The SPI and I2C peripherals allow the DSPs to compute in software functions such as automatic gain control (AGC) and send the information seamlessly to the RF control interface. The DSP computes the changed values and the SPI or I2C bus delivers the information to the external chip(s).

2.5. Power management

To facilitate flexible system-level power management, the Sandblaster SB3011 incorporates thirteen independent power domains. Each processor core is isolated by a separate domain thus 5 domains encapsulate the ARM plus 4 DSPs. An additional domain is used for L2 memories. The other power domains are used to isolate specific logic portions of the chip.

Each domain is independently controllable by the Device Power Management Unit (DPMU) which is itself isolated within an independent power domain. The DPMU is a programmable peripheral that allows for the following options: (1) the ability to place the device in power down where all data and internal state is not maintained and (2) the ability to place each processor independently in power down where each core does not maintain state but the L2 memories are back-biased and thus retain state.

In addition to the voltage control features, clock management is also included in two forms: (1) instruction-based automatic clock enable/disable operation where the hardware dynamically controls clocks to internal sub-blocks when inactivity is detected and (2) operating System (OS) or application-based clock enable/disable which can control DSP cores, AHB peripherals, LCD, Camera, USB, Ethernet, and APB peripherals.

While not a comprehensive list, some typical profiles of low power configurations include the following. (1) Device Deep Sleep where the all the SB3011 functional blocks are powered off with the exception of the Device Power Management Unit. No state is retained in this mode. In this state only the DPMU is powered since it is required to wake the rest of the chip. (2) Processing Unit Deep Sleep Mode where all the processor cores are shut down without state retention. However, L2 memories and peripherals retain state and may function. (3) Device Standby where all DSP cores and the ARM processor clocks are disabled but full state is retained.

The subsequent sections discuss the Sandblaster DSP architecture and programming tools that enable real-time implementation of the parallel SDR system.

3. SANDBLASTER LOW-POWER ARCHITECTURE

3.1. Compound instructions

The Sandblaster architecture is a compound instruction set architecture [7]. Historically, DSPs have used compound

```
L0: lvu %vr0, %r3, 8

||                      vmulreds
%ac0,%vr0,%vr0,%ac0

|| loop %lc0,L0
```

FIGURE 2: Compound instruction for sum of squares inner loop.

instruction set architectures to conserve instruction space encoding bits. In contrast, some VLIW architectures contain full orthogonality, but only encode a single operation per instruction field, such that a single VLIW is composed of multiple instruction fields. This has the disadvantage of requiring many instruction bits to be fetched per cycle, as well as significant write ports for register files. Both these effects contribute heavily to power dissipation. Recent VLIW DSP architectures, such as STA, overcome these limitations by providing complex operations, vector operations, and multiple register files.

In the Sandblaster architecture, specific fields within the instruction format may issue multiple suboperations including data parallel vector operations. Most classical DSP instruction set architectures are compound but impose restrictions depending upon the particular operations selected. In contrast, some VLIW ISAs allow complete orthogonality of specification and then fill in any unused issue slots by inserting no operation instructions (NOPs) either in hardware or software.

3.2. Vector encoding

In addition to compound instructions, the Sandblaster architecture also contains vector operations that perform multiple compound operations. As an example, Figure 2 shows a single compound instruction with three compound operations. The first compound operation, lvu, loads the vector register vr0 with four 16-bit elements and updates the address pointer r3 to the next element. The vmulreds operation reads four fixed point (fractional) 16-bit elements from vr0, multiplies each element by itself, saturates each product, adds all four saturated products plus an accumulator register, ac0, with saturation after each addition, and stores the result back in ac0. The vector architecture guarantees Global System for Mobile communication (GSM) semantics (e.g., bit-exact results) even though the arithmetic performed is nonassociative [8]. The loop operation decrements the loop count register lc0, compares it to zero, and branches to address L0 if the result is not zero.

3.3. Simple instruction formats

Simple and orthogonal instruction formats are used for all instructions. The type of operation is encoded to allow simple decoding and execution unit control. Multiple operation fields are grouped within the same bit locations. All operand fields within an operation are uniformly placed in the same

bit locations whether they are register-based or immediate values. As in VLIW processors, this significantly simplifies the decoding logic.

3.4. Low-power idle instructions

Architecturally, it is possible to turn off an entire processor. All clocks may be disabled or the processor may idle with clocks running. Each hardware thread unit may also be disabled to minimize toggling of transistors in the processor.

3.5. Fully interlocked

Unlike some VLIW processors, our architecture is fully interlocked and transparent. In addition to the benefit of code compatibility, this ensures that many admissible and application-dependent implementations may be derived from the same basic architecture.

4. LOW-POWER MICROARCHITECTURE

4.1. Multithreading

Figure 3 shows the microarchitecture of the Sandblaster processor. In a multithreaded processor, multiple threads of execution operate simultaneously. An important point is that multiple copies (e.g., banks and/or modules) of memory are available for each thread to access. The Sandblaster architecture supports multiple concurrent program execution by the use of hardware thread units (called contexts). The architecture supports up to eight concurrent hardware contexts. The architecture also supports multiple operations being issued from each context. The Sandblaster processor uses a new form of multithreading called token triggered threading (T^3) [9].

With T^3, all hardware contexts may be simultaneously executing instructions, but only one context may issue an instruction each cycle. This constraint is also imposed on round-robin threading. What distinguishes T^3 is that each clock cycle, a token indicates the next context that is to be executed. Tokens may cause the order in which threads issue instructions to be sequential (e.g., round-robin), even/odd, or based on other communications patterns. Figure 4 shows an example of T^3 instruction issue, in which an instruction first issues from Thread 0, then Thread 3, then Thread 2, and so forth. After eight cycles, the sequence repeats with Thread 0 issuing its next instruction. Compared to SMT, T^3 has much less hardware complexity and power dissipation, since the method for selecting threads is simplified, only a single compound instruction issues each clock cycle, and most dependency checking and bypass hardware is not needed.

4.2. Decoupled logic and memory

As technology improves, processors are capable of executing at very fast cycle times. Current state-of-the-art 0.13 um technologies can produce processors faster than 3 GHz. Unfortunately, current high-performance processors consume

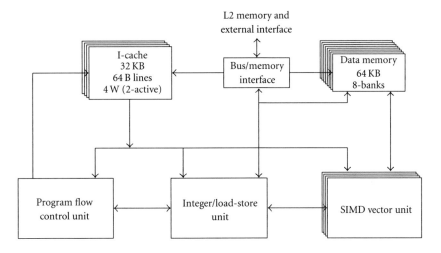

FIGURE 3: Multithreaded microarchitecture.

→ T0 → T3 → T2 → T1 → T6 → T5 → T4 → T7 →

FIGURE 4: Token triggered threading, with even/odd sequencing.

significant power. If power-performance curves are considered for both memory and logic within a technology, there is a region in which you get approximately linear increase in power for linear increase in performance. Above a specific threshold, there is an exponential increase in power for a linear increase in performance. Even more significant, memory and logic do not have the same threshold.

For 0.13 um technology, the logic power-performance curve may be in the linear range until approximately 600 MHz. Unfortunately, memory power-performance curves are at best linear to about 300 MHz. This presents a dilemma as to whether to optimize for performance or power. Fortunately, multithreading alleviates the power-performance trade-off. The Sandblaster implementation of multithreading allows the processor cycle time to be decoupled from the memory access time. This allows both logic and memory to operate in the linear region, thereby significantly reducing power dissipation. The decoupled execution does not induce pipeline stalls due to the unique pipeline design.

4.3. Caches

An instruction cache unit (ICU) stores instructions to be fetched for each thread unit. A cache memory works on the principle of locality. Locality can refer to spatial, temporal, or sequential locality [2]. We use set associative caches to alleviate multiple contexts evicting another context's active program. In our implementation, shown in Figure 5, there are four directory entries (D0–D3) and banked storage entries. A thread identifier register (not shown) is used to se-

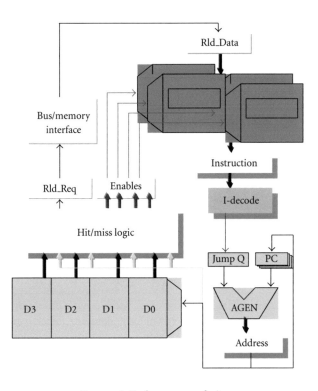

FIGURE 5: Cache memory design.

lect whether the cache line in the left or right bank will be evicted. This effectively reduces the complexity of the cache line selection logic. In a 4-way set associative cache, only one additional least recently used (LRU) bit is needed to select which of the two lines should be evicted. This method of using thread information and banked memory accesses significantly reduces the complexity of the cache logic. In our implementation, a unique feature is the use of a read associativity of 4 and a write associativity of 2, which further reduces the cache logic complexity.

Ld/St	Inst. Dec.	RF read	Agen.	XFer	Int. ext.	Mem. 0	Mem. 1	Mem. 2	WB
ALU	Inst. Dec.	Wait	RF read	Exec. 1	Exec. 2	XFer	WB	—	—
I_Mul	Inst. Dec.	Wait	RF read	Exec. 1	Exec. 2	Exec. 3	XFer	WB	—
V_Mul	Inst. Dec.	VRF read	Mpy1	Mpy2	Add1	Add2	XFer	VRF WB	—

FIGURE 6: Processor pipeline.

4.4. Pipeline

The pipeline for one particular implementation of the Sand-blaster DSP is shown in Figure 6. The execution pipelines are different for various functions. The Load/Store (Ld/St) pipeline is shown to have 9 stages. It is assumed that the instruction is already in the cache. The first stage decodes the instruction (Inst. Dec.). This is followed by a read from the general purpose register file. The next stage generates the address to perform the Load or Store. Five cycles are used to access data memory. Finally, the result for a Load instruction is written back (WB) to the referenced register file location. Once an instruction from a particular context enters the pipeline, it runs to completion. It is also guaranteed to write back its result before the next instruction issuing from the same thread tries to use the result.

Similarly, there are multiple (variable) stages for other execution pipelines. The integer unit has three execute stages for multiplication (I_MUL) and two execute stages for addition (ALU). The vector unit has four execute stages, two for multiplication and two for addition.

4.5. Interlock checking hardware

Most interlocked architectures require significant interlock checking hardware and bypass logic for both correctness and performance reasons. Multithreading mitigates this effect. With the carefully designed pipeline shown in Figure 6, there is only one interlock that must actually be checked for in hardware, a long memory load or store. All other operations are guaranteed to complete prior to the same thread issuing a new instruction. This completely removes the power consuming interlock checks associated with most interlocked architectures.

5. LOW-POWER LOGIC DESIGN

5.1. Single write-port register files

Having multithreading to cover the latency associated with long pipeline implementations allows the use of single write-port register files even though more than one write may occur within an instruction cycle. An important point is that the write back stages are staggered. This allows a single write port to be implemented but provides the same functionality as multiple write ports [10].

An example is loading the integer register file while performing an integer multiply. From the processor pipeline shown in Figure 6, it is apparent that the reads and writes from the register file are staggered in time. In addition, separate architected register spaces for vector, integer, and accumulate operations enable reduced ports. A VLIW implementation of the instruction shown in Figure 2 may take many write ports for sustained single cycle throughput. Comparatively, our solution requires at most a single combined R/W port and an additional read port per register file.

5.2. Banked register files

Token triggered threading which follows a permutation of even and odd thread issue policies along with the pipeline implementation enables the use of banked register files. This allows the register files to run at half the processor clock, but never stall awaiting data.

5.3. Single-ported memories

The same characteristics that allow banked register file operation also enable the use of single ported L1 memories that may also be banked and run at half the processor clock. Since decoupled memories are highly desirable to reduce power, this provides significant overall savings.

5.4. Minimal control signals

A combination of architectural and microarchitectural techniques allows the processor to be implemented with very few control signals. Since control signals often propagate to many units, they become not only a source of bugs but also may dissipate significant power.

5.5. Clock gating

Because the architecture is modular and the pipeline is deep, there is time to compute which functional units will be active for each instruction. If a functional unit is not active, the clocks may be gated to that unit and suspend it on a unit-by-unit basis. As an example, if there are no vector operations on a given cycle, the vector unit is disabled. Even within a unit it is possible to gate the clocks. For example, if a vector multiply operation is executed but it does not need to be reduced, the reduce unit within the vector unit is gated off.

6. LOW-POWER CIRCUIT DESIGN

The average power consumption in a CMOS circuit can be modeled as

$$P_{\text{avg}} = \alpha C V_{\text{dd}}^2 f + V_{\text{dd}} I_{\text{mean}} + V_{dd} I_{\text{leak}}, \qquad (1)$$

where α is the average gate switching activity, C is the total capacitance seen by the gates' outputs, V_{dd} is the supply voltage, f is the circuit's operating frequency, I_{mean} is the average current drawn during input transition, and I_{leak} is the average leakage current. The first term, $\alpha C V_{\text{dd}}^2 f$, which represents the dynamic switching power consumed by charging and discharging the capacitive load on the gates' outputs, often dominates power consumption in high-speed microprocessors [11]. The second term, $V_{\text{dd}} I_{\text{mean}}$, which represents the average dynamic power due to short-circuit current flowing when both the PMOS and NMOS transistors conduct during input signal transitions, typically contributes 10% to 20% of the overall dynamic power [12]. This is also a function of frequency but is simplified in this analysis. The third term, $V_{\text{dd}} I_{\text{leak}}$, represents the power consumed due to leakage current and occurs even in devices that are not switching. Consequently, for systems that are frequently in standby mode, the leakage power may be a dominate factor in determining the overall battery life. Since the leakage power increases exponentially with a linear decrease in device threshold voltage, leakage power is also a concern in systems that use power supply voltage scaling to reduce power.

6.1. Low-voltage operation

Since the dynamic switching power, $\alpha C V_{\text{dd}}^2 f$, is proportional to the supply voltage squared, an effective technique for reducing power consumption is to use a lower supply voltage. Unfortunately, however, decreasing the supply voltage also decreases the maximum operating frequency. To achieve high performance with a low-supply voltage, our arithmetic circuits are heavily pipelined. For example, our multiply-accumulate unit uses four pipeline stages. Our unique form of multithreading helps mask long pipeline latencies, so that high performance is achieved.

6.2. Minimum dimension transistors

Minimum dimension transistors help to further reduce power consumption, since they reduce circuit capacitance [13]. Throughout the processor, we use minimum dimension transistors, unless other considerations preclude their use. For example, transistors that are on critical delay paths often need to have larger dimensions to reduce delay [14]. Larger dimension transistors are also used to drive nodes with high fan-out and to balance circuit delays.

6.3. Delay balancing

Gates with unbalanced input delays can experience glitches, which increase dynamic switching power and dynamic short-circuit power [15]. To reduce glitches, we balance gate input delays in our circuits through a combination of gate-level delay balancing techniques (i.e., designing the circuits so that inputs to a particular gate go through roughly the same number of logic levels) and judicious transistor sizing. Glitches are further reduced by having a relatively small number of logic levels between pipeline registers.

6.4. Logic combining and input ordering

Dynamic and static power consumptions are also reduced by utilizing a variety of specially designed complex logic cells. Our circuits include efficient complex logic cells, such as 3-input AndOrInvert (AOI), 3-input OrAndInvert (OAI), half adder, and full adder cells. Providing a wide variety of complex gates with different drive strengths, functionality, and optionally inverted inputs gives circuit designers and synthesis tools greater flexibility to optimize for power consumption. Keeping nodes with a high probability of switching inside of complex gates and reordering the inputs to complex gates can help further reduce power. In general, inputs that are likely to be off are placed closer to gate output nodes, while inputs that are likely to be on are placed closer to the supply voltage [15].

7. SANDBLASTER SOFTWARE TOOLS

A *simulator* is an interpreter of a machine language where the representation of programs resides in memory but is not directly executed by host hardware. Historically, three types of architectural simulators have been identified. An *interpreted simulator* consists of a program executing on a computer where each machine language instruction is executed on a model of a target architecture running on the host computer. Because interpreted simulators tend to execute slowly, compiled simulators have been developed. A *statically compiled simulator* first translates both the program and the architecture model into the host computer's machine language. A *dynamically compiled* (or just-in-time) *simulator* either starts execution as an interpreter, but judiciously chooses functions that may be translated during execution into a directly executable host program, or begins by translating at the start of the host execution.

7.1. Interpreted execution

Instructions set simulators commonly used for application code development are cycle-count accurate in nature. They use an architecture description of the underlying processor and provide close-to-accurate cycle counts, but typically do not model external memories, peripherals, or asynchronous interrupts. However, the information provided by them is generally sufficient to develop the prototype application.

Figure 7 shows an interpreted simulation system. Executable code is generated for a target platform. During the execution phase, a software interpreter running on the host interprets (simulates) the target platform executable. The simulator models the target architecture, may mimic the implementation pipeline, and has data structures to reflect the

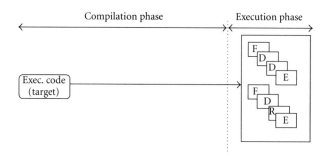

FIGURE 7: Interpreted simulation.

machine resources such as registers. The simulator contains a main driver loop, which performs the *fetch, decode, data read, execute, and write back* operations for each instruction in the target executable code.

An interpreted simulator has performance limitations. Actions such as instruction fetch, decode, and operand fetch are repeated for every execution of the target instruction. The instruction decode is implemented with a number of conditional statements within the main driver loop of the simulator. This adds significant simulation overhead because all combinations of opcodes and operands must be distinguished. In addition, the execution of the target instruction requires the update of several data structures that mimic the target resources, such as registers, in the simulator.

7.2. Statically compiled simulation

Figure 8 shows a statically compiled simulation system. In this technique, the simulator takes advantage of the any a priori knowledge of the target executable and performs some of the activities at compile time instead of execution time. Using this approach, a simulation compiler generates host code for instruction fetch, decode, and operand reads at compile time. As an end product, it generates an application-specific host binary in which only the execute phase of the target processor is unresolved at compile time. This binary is expected to execute faster, as repetitive actions have been taken care of at compile time.

While this approach addresses some of the issues with interpretive simulators, there are other limitations. First, the simulation compilers typically generate C code, which is then converted to object code using the standard *compile → assemble → link* path. Depending on the size of the generated C code, the file I/O needed to scan and parse the program could well reduce the benefits gained by taking the compiled simulation approach. The approach is also limited by the idiosyncrasies of the host compiler such as the number of labels allowed in a source file, size of switch statements and so forth. Some of these could be addressed by directly generating object code—however, the overhead of writing the application-specific executable file to the disc and then rereading it during the execution phase still exists. In addition, depending on the underlying host, the application-specific executable (which is visible to the user) may not be portable to another host due to different libraries, instruction sets and so forth.

7.3. Dynamically compiled simulation

Figure 9 shows the dynamically compiled simulation approach. This is the approach used in the Sandbridge simulator. In this approach, target instructions are translated into equivalent host instructions (executable code) at the beginning of execution time. The host instructions are then executed at the end of the translation phase. This approach eliminates the overhead of repetitive target instruction fetch, decode, and operand read in the interpretive simulation model. By directly generating host executable code, it eliminates the overhead of the compile, assemble, and link path and the associated file I/O that is present in the compiled simulation approach. This approach also ensures that the target executable file remains portable, as it is the only executable file visible to the user and the responsibility of converting it to host binary has been transferred to the simulator.

7.4. Multithreaded programming model

Obtaining full utilization of parallel processor resources has historically been a difficult challenge. Much of the programming effort can be spent determining which processors should receive data from other processors. Often execution cycles may be wasted for data transfers. Statically scheduled machines such as Very Long Instruction Word architectures and visible pipeline machines with wide execution resources complicate programming and may reduce programmer productivity by requiring manual tracking of up to 100 in-flight instruction dependencies. When nonassociative DSP arithmetic is present, nearly all compilers are ineffective and the resulting burden falls upon the assembly language programmer. A number of these issues have been discussed in [8].

A good programming model should adequately abstract most of the programming complexity so that 20% of the effort may result in 80% of the platform utilization [16]. While there are still some objections to a multithreaded programming model [9], to-date it is widely adopted particularly with the introduction of the Java programming language [17].

With hardware that is multithreaded with concurrent execution and adopting a multithreaded software programming model, it is possible for a kernel to be developed that automatically schedules software threads onto hardware threads. It should be noted that while the hardware scheduling is fixed and uses a technique called token triggered threading (T^3) [18], the software is free to use any scheduling policy desired.

The Sandblaster kernel has been designed to use the POSIX pthreads open standard [19]. This provides cross platform capability as the library is compilable across a number of systems including Unix, Linux, and Windows.

7.5. Compiler technology

There are many challenges faced when trying to develop efficient compilers for parallel DSP technologies. At each level of processor design, Sandbridge has endeavored to alleviate these issues through abstraction. First and foremost,

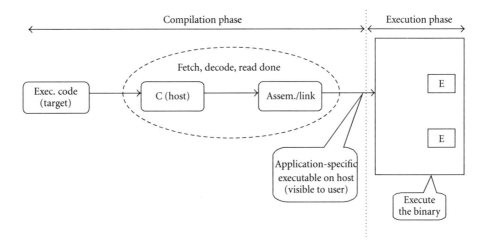

FIGURE 8: Statically compiled simulation.

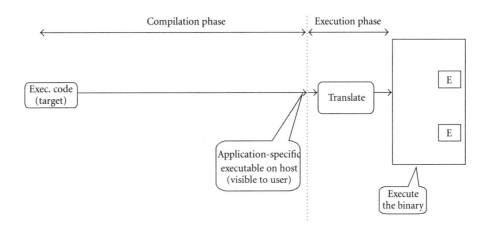

FIGURE 9: Dynamically compiled simulation.

the Sandblaster processor is transparent in the architectural sense. This proscribes that there are no visible implementation effects for the programmer or compiler to deal with [2]. This is in distinct contrast with VLIW designs where the implementation strongly influences the architecture. A benefit of a true architecture approach is that object code will execute unmodified (e.g., without any translation required) on any Sandblaster compliant implementation.

The Sandblaster architecture uses a SIMD datapath to implement vector operations. The compiler vectorizes C code to exploit the data level parallelism inherent in signal processing applications and then generates the appropriate vector instructions. The compiler also handles the difficult problem of outer loop vectorization

Within the architecture, there is direct support for parallel saturating arithmetic. Since saturating arithmetic is nonassociative, out-of-order execution may produce different bit results. In some wireless systems this is not permissible [20]. By architecting parallel saturating arithmetic (i.e., vector multiply and accumulate with saturation), the compiler is able to generate code with the understanding that the

hardware will properly produce bit-exact results. The compiler algorithm used to accomplish this is described in [21]. Some hardware techniques to implement this are described in [22].

Additionally, our compiler can also automatically generate threads. We use the same pthreads mechanism for thread generation in the compiler as the programmer who specifies them manually. For most signal processing loops, it is not a problem to generate threads and the compiler will automatically produce code for correct synchronization.

7.6. Tool chain generation

Figure 10 shows the Sandblaster tool chain generation. The platform is programmed in a high-level language such as C, C++, or Java. The program is then translated using an internally developed supercomputer class vectorizing parallelizing compiler. The tools are driven by a parameterized resource model of the architecture that may be programmatically generated for a variety of implementations and organizations. The source input to the tools, called the Sandbridge

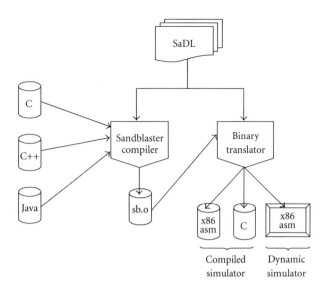

FIGURE 10: Tool chain generation.

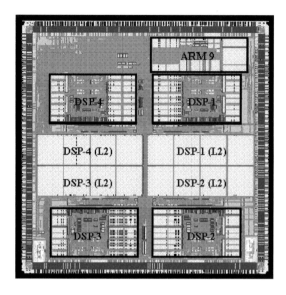

FIGURE 11: SB3011 device layout.

architecture description language (SaDL), is a collection of python source files that guide the generation and optimization of the input program and simulator. The compiler is retargetable in the sense that it is able to handle multiple possible implementations specified in SaDL and produce an object file for each implementation. The platform also supports many standard libraries (e.g., libc, math, etc.) that may be referenced by the C program. The compiler generates an object file optimized for the Sandblaster architecture.

8. RESULTS

This section discusses the performance and power results for the processor, the simulation and compilation performance results, and finally full communications systems results.

8.1. Processor performance and power results

Figure 11 shows a picture of the SB3011 chip which was fabricated in 90 nm TSMC technology. Highlighted are the 4 Sandblaster cores, the ARM9 core, and the L2 memories. Initial samples have performed at 600 MHz at 0.9 V.

Figure 12 shows power measurements made on the initial samples for a single Sandblaster core. As described in Section 2.5, the power modes may be programmed. Figure 12 shows power at some typical configurations. When the entire device is in deep sleep it consumes less than 1 microwatt of power. As you bring each core out of deep sleep to a standby state, there is a measured range of power dissipation which on the initial samples is less than 5 milliwatts with complete state retention. The last section of Figure 12 depicts the linear nature of programs executing. Depending on the core activity, the power dissipation is linear with respect to the workload. The linear nature depicted is the result of average utilization of threads. We have measured on hardware a range of applications. WCDMA dissipates about 75 mW per

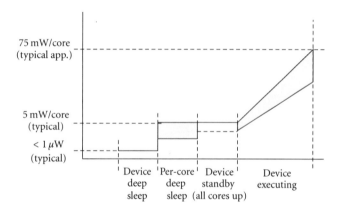

FIGURE 12: Processor power results for a 600 MHz 0.9 V Sandblaster device.

core (at 600 MHz 0.9 V). Other less demanding applications such as GSM/GPRS dissipate less power.

8.2. Processor tools results

Figure 13 shows the results of various compilers on out-of-the-box ETSI C code [20]. The y-axis shows the number of MHz required to compute frames of speech in real-time. The AMR code is completely unmodified and no special include files are used. Without using any compiler techniques such as intrinsics or special typedefs, the compiler is able to achieve real-time operation on the baseband core at hand-coded assembly language performance levels. Note that the program is completely compiled from C language code. Since other solutions are not able to automatically generate DSP operations, intrinsic libraries must be used. With intrinsic libraries the results for most DSPs are near ours but they only apply to the ETSI algorithms whereas the described compiler can be applied to arbitrary C code.

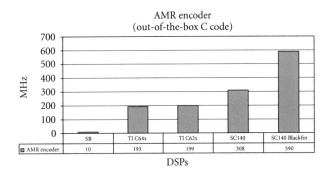

FIGURE 13: Out-of-the-box AMR ETSI encoder C code results. (Results based on out-of-the-box C code. C64x IDE Version 2.0.0 compiled without intrinsics using -k -q -pm -op2 -o3 -d"WMOPS = 0" -ml0 -mv6400 flags with results averaged over 425 frames of ETSI-supplied test vectors. C62x IDE Version 2.0.0 compiled without intrinsics using -k -q -pm -op2 -o3 -d"WMOPS = 0" -ml0 -mv6200 flags with results averaged over 425 frames of ETSI-supplied test vectors. Starcore SC140 IDE version Code Warrior for StarCore version 1.5, relevant optimization flags (encoder only): scc -g -ge -be -mb -sc -O3 –Og, other: no intrinsic used. Results based on execution of 5 frames. ADI Blackfin IDE Version 2.0 and Compiler version 6.1.5 compiled without intrinsics using -O1 -ipa -DWMOPS = 0 –BLACKFIN with results averaged over 5 frames of ETSI-supplied test vectors for the encoder only portion.)

Efficient compilation is just one aspect of software productivity. Prior to having hardware, algorithm designers should have access to fast simulation technology. Figure 14 shows the postcompilation simulation performance of the same AMR encoder as Figure 13 for a number of DSP processors. All programs were executed on the same 1 GHz laptop Pentium computer. The Sandbridge tools are capable of simulating 24.6 million instructions per second. This is more than two orders of magnitude faster than the nearest DSP and allows real-time execution of GSM speech coding on a Pentium simulation model. To further elaborate, while some DSPs cannot even execute the out-of-the-box code in real-time on their native processor, the Sandbridge simulator achieves multiple real-time channels on a simulation model of the processor. This was accomplished by using internal compilation technology to accelerate the simulation.

8.3. Applications results

Figure 15 shows the results of a number of communications systems as a percentage utilization of a 4-core 600 MHz SB3011 platform. Particularly, WiFi 802.11b, GPS, AM/FM radio, Analog NTSC Video TV, Bluetooth, GSM/GPRS, UMTS WCDMA, WiMax, CDMA, and DVB-H. A notable point is that all these communications systems are written in generic C code with no hardware acceleration required. It is also notable that performance in terms of data rates and concurrency in terms of applications can be dynamically adjusted based on the mix of tasks desired. For most of the systems, the values are measured on hardware from digitized RF signals that have been converted in real-time. This includes

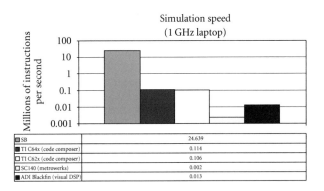

FIGURE 14: Simulation speed of ETSI AMR encoder.

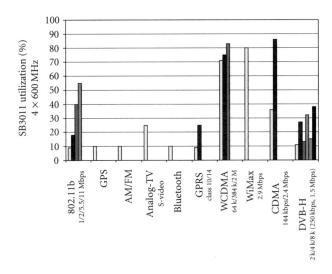

FIGURE 15: Communication systems results as a percentage of SB3011 utilization (4 Cores at 600 MHz).

the design of RF cards based on industry standard components. The only exceptions are Bluetooth and DVB-H. For these systems the RF cards are still under development.

Figure 16 shows the results of various multimedia codecs. Note that the total MHz scale is 7% of the entire 4-core capacity. Results for QCIF (176 × 144) at 15 frames per second (fps) and CIF (360 × 288) at 30 fps images are shown for the H.264 decoder. For the MPEG4 decoder, out-of-the-box (OOB) and optimized (OPT) are shown for the Foreman clip at 30 frames per second. Noticeably, out-of-the-box performance is real-time and highly efficient. This is the result of our highly optimizing compiler which can both vectorize and parallelize standard C code. Also, MP3 decoding results are shown at various bit rates. A key point is that all these applications run using less than two threads and many in a percentage of a single thread. Since there are 32 threads in the SB3011 implementation, a single thread consumes 3.125% of the available processor performance.

Figure 17 shows measurements while executing either GPRS class 14 or WCDMA at a 384 kbps bit rate. Note that both of these applications dissipate less power than the stated average dissipation of 75 mW per core. The actual power

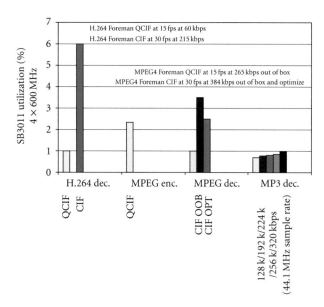

FIGURE 16: Multimedia results as a percentage of SB3011 utilization (4 cores at 600 MHz).

dissipation is highly dependent upon workload. As an approximation it may be possible to use the average utilization of the processor complex multiplied by the average power. However, in practice the actual results vary significantly by application. The SB3010 measurements refer to an earlier version of the chip that was predominantly synthesized. The SB3011 is a semicustom design. The software-optimized column refers to the operating system's ability to turn off cores and threads that are unused. This can result in significant power savings.

9. RELATED WORK

In this section we contrast and compare our approach for both processors and tools with other known approaches. The Sandbridge processor design provides important power and performance characteristics while the tools provide the capability of rapidly designing SDR systems.

9.1. Processors

Other SDR platforms include the Signal Processing on Demand Architecture (SODA) [23], OnDSP [4], the Embedded Vector Processor (EVP) [5], the Synchronous Transfer Architecture (STA) [6], picoArray [24], XiSystem [25], and the MS1 reconfigurable DSP (rDSP) core [26].

SODA is a programmable SDR platform that consists of four processor cores. Each core contains scratchpad memories and asymmetric pipelines that support scalar, 32-wide SIMD, and address generation operations. SODA is optimized for 16-bit arithmetic and features several specialized operations including saturating arithmetic, vector permute, vector compare and select, and predicated negation operations.

OnDSP, EVP, and STA all are VLIW architectures with support for multiple parallel scalar, vector, memory access, and control operations. For example, OnDSP provides 8-element vector operations that can operate in parallel with scalar operations. With EVP, the maximum VLIW-parallelism available is five vector operations, four scalar operations, three address updates, and loop-control. All three architectures feature dedicated instructions for wireless communications algorithms, such as FFTs and Viterbi, Reed-Solomon, and Turbo coding. STA utilizes a machine description file to facilitate the generation of different hardware and simulation models for the processor.

picoArray is a tiled architecture in which hundreds of heterogeneous processors are interconnected using a bus-based array. Within the picoArray, processors are organized in a two-dimensional grid, and communicate over a network of 32-bit unidirectional buses and programmable bus switches. Each programmable processor in the array supports 16-bit arithmetic, uses 3-way VLIW scheduling, and has its own local memory. In addition to the programmable processors, the picoArray includes specialized peripherals and connects to hardware accelerators for performing FFTs, cryptography, and Reed-Solomon and Viterbi coding.

XiSystem and the MS1 rDSP core combine programmable processors with reconfigurable logic to implement wireless communication systems. XiSystem integrates a VLIW processor, a multicontext reconfigurable gate array, and reconfigurable I/O modules in a SoC platform. The multicontext reconfigurable gate array enables dynamic instruction set extensions for bit-level operations needed in many DSP applications. The MS1 rDSP core contains a reconfigurable logic block, called the RC Array, a 32-bit RISC processor, called mRISC, a context memory, a data buffer, and an I/O controller. The mRISC processor controls the RC array, which performs general purpose operations, as well as word-level and bit-level DSP functions.

Unlike other SDR platforms, the SB3011 platform provides a fully programmable solution in which all communications systems are written in generic C code with no hardware acceleration or assembly language programming. It also is the first SDR platform to combine explicit multithreading, powerful compound instructions, vector operations, and parallel saturating arithmetic in a low-power programmable SoC multiprocessor.

Another important aspect of the SB3011 platform is the technique it uses to support explicit multithreading. Previous techniques for explicit hardware multithreading including interleaved multithreading (IMT), blocked multithreading (BMT), and simultaneous multithreading (SMT) [27]. With IMT [28], also known as fine grain multithreading or horizontal multithreading, only one thread can issue an instruction each cycle, and threads issue instructions in a predetermined order (e.g., round-robin). With BMT [29], also known as coarse-grain multithreading or vertical multithreading, instructions are executed sequentially until a long-latency event (e.g., a cache miss) occurs. The long-latency event triggers a fast context switch to another thread. With SMT [30], multiple instructions may be issued each cycle

	GSM/GPRS class 14, 1 core			WCDMA at 384 kbps, 1 of 3 cores		
	SB3010 (mW)	SB3010 (mW) SW optimized	SB3011 (mW)	SB3010 (mW)	SB3010 (mW) SW optimized	SB3011 (mW)
	Measured			Measured		
Core w/ L1 instances 32 KB I-cache 64 KB D mem.	150	85	45	171	130	65
Core w/o L1 instances	142	77	40	160	117	58

FIGURE 17: Application power measurements.

from multiple threads. SMT combines techniques from previous multithreaded processors and dynamically scheduled superscalar processors to exploit both instruction-level parallelism and thread-level parallelism.

As discussed in Section 4, The SB3011 features a new form of interleaved multithreading, known as token triggered threading (T^3). Unlike previous IMT implementations, the T^3 implementation on the SB3011 features compound instructions, SIMD vector operations, and greater flexibility in scheduling threads. Compared to BMT, T^3 provides greater concurrency since instructions from multiple threads are executing in parallel each cycle. Compared to SMT, T^3 has much less hardware complexity and power dissipation, since the method for selecting threads is simplified, only a single compound instruction issues each clock cycle, and dependency checking and bypass hardware are not needed. The SB3011 platform combines T^3 with chip multiprocessing to provide up to 32 simultaneously executing hardware threads.

9.2. Tools

In this section, we compare our solution to other high-performance tools solutions. Automatic DSP simulation generation from a C++-based class library was discussed in [31]. Automatic generation of both compiled and interpretive simulators was discussed in [32]. Compiled simulation for programmable DSP architectures to increase simulation performance was introduced in [33]. This was extended to cycle accurate models of pipelined processors in [34]. A general purpose MIPS simulator was discussed in [35]. The ability to dynamically translate snippets of target code to host code at execution time was used in Shade [36]. However, unlike Shade, our approach generates code for the entire application, is targeted towards compound instruction set architectures, and is capable of maintaining bit exact semantics of DSP algorithms. A similar approach to ours is described in [37].

10. SUMMARY

Sandbridge Technologies has introduced a completely new and scalable design methodology for implementing multiple communications systems on a single SDR chip. Using a unique multithreaded architecture specifically designed to reduce power consumption, efficient broadband communications operations are executed on a programmable plat-

form. The instruction execution in the described architecture is completely interlocked providing software compatibility among all processors. Because of the interlocked execution, interrupt latency is very short. An interrupt may occur on any instruction boundary including loads and stores; this is critical for real-time systems.

The processor is combined with a highly optimizing vectorizing compiler with the ability to automatically analyze programs and generate DSP instructions. The compiler also automatically parallelizes and multithreads programs. This obviates the need for assembly language programming and significantly accelerates time-to-market for streaming multimode multimedia convergence systems.

REFERENCES

[1] http://www.sdrforum.org/.

[2] G. Blaauw and F. Brooks Jr., *Computer Architecture: Concepts and Evolution*, Addison-Wesley, Reading, Mass, USA, 1997.

[3] B. Case, "Philips hopes to displace DSPs with VLIW," *Microprocessor Report*, vol. 8, no. 16, pp. 12–15, 1997.

[4] J. Kneip, M. Weiss, W. Drescher, et al., "Single chip programmable baseband ASSP for 5 GHz wireless LAN applications," *IEICE Transactions on Electronics*, vol. E85-C, no. 2, pp. 359–367, 2002.

[5] K. van Berkel, F. Heinle, P. P. E. Meuwissen, K. Moerman, and M. Weiss, "Vector processing as an enabler for software-defined radio in handheld devices," *EURASIP Journal on Applied Signal Processing*, vol. 2005, no. 16, pp. 2613–2625, 2005.

[6] J. P. Robelly, G. Cichon, H. Seidel, and G. Fettweis, "A HW/SW design methodology for embedded SIMD vector signal processors," *International Journal of Embedded Systems*, vol. 1, no. 11, pp. 2–10, 2005.

[7] J. Glossner, T. Raja, E. Hokenek, and M. Moudgill, "A multi-threaded processor architecture for SDR," *The Proceedings of the Korean Institute of Communication Sciences*, vol. 19, no. 11, pp. 70–84, 2002.

[8] J. Glossner, M. Schulte, M. Moudgill, et al., "Sandblaster low-power multithreaded SDR baseband processor," in *Proceedings of the 3rd Workshop on Applications Specific Processors (WASP '04)*, pp. 53–58, Stockholm, Sweden, September 2004.

[9] E. A. Lee, "The problem with threads," *Computer*, vol. 39, no. 5, pp. 33–42, 2006.

[10] J. Glossner, K. Chirca, M. Schulte, et al., "Sandblaster low power DSP," in *Proceedings of the IEEE Custom Integrated Circuits Conference (CICC '04)*, pp. 575–581, Orlando, Fla, USA, October 2004.

[11] B. Moyer, "Low-power design for embedded processors," *Proceedings of the IEEE*, vol. 89, no. 11, pp. 1576–1587, 2001.

[12] T. Mudge, "Power: a first-class architectural design constraint," *Computer*, vol. 34, no. 4, pp. 52–58, 2001.

[13] A. Wroblewski, O. Schumacher, C. V. Schimpfle, and J. A. Nossek, "Minimizing gate capacitances with transistor sizing," in *Proceedings of the IEEE International Symposium on Circuits and Systems (ISCAS '01)*, vol. 4, pp. 186–189, Sydney, NSW, Australia, May 2001.

[14] M. Borah, R. M. Owens, and M. J. Irwin, "Transistor sizing for minimizing power consumption of CMOS circuits under delay constraint," in *Proceedings of the International Symposium on Low Power Electronics and Design*, pp. 167–172, Dana Point, Calif, USA, April 1995.

[15] S. Kim, J. Kim, and S.-Y. Hwang, "New path balancing algorithm for glitch power reduction," *IEE Proceedings: Circuits, Devices and Systems*, vol. 148, no. 3, pp. 151–156, 2001.

[16] R. Goering, "Platform-based design: a choice, not a panacea," *EE Times*, 2002, http://www.eetimes.com/story/OEG2002091-1S0061.

[17] O. Silvén and K. Jyrkkä, "Observations on power-efficiency trends in mobile communication devices," in *Proceedings of the 5th International Workshop on Embedded Computer Systems: Architectures, Modeling, and Simulation (SAMOS '05)*, vol. 3553 of *Lecture Notes in Computer Science*, pp. 142–151, Samos, Greece, July 2005.

[18] M. Schulte, J. Glossner, S. Mamidi, M. Moudgill, and S. Vassiliadis, "A low-power multithreaded processor for baseband communication systems," in *Embedded Processor Design Challenges: Systems, Architectures, Modeling, and Simulation*, vol. 3133 of *Lecture Notes in Computer Science*, pp. 393–402, Springer, New York, NY, USA, 2004.

[19] B. Nichols, D. Buttlar, and J. Farrell, *Pthreads Programming: A POSIX Standard for Better Multiprocessing*, O'Reilly Nutshell Series, O'Reilly Media, Sebastopol, Calif, USA, 1996.

[20] K. Jarvinen, J. Vainio, P. Kapanen, et al., "GSM enhanced full rate speech codec," in *Proceedings of IEEE International Conference on Acoustics, Speech and Signal Processing (ICASSP '97)*, vol. 2, pp. 771–774, Munich, Germany, April 1997.

[21] V. Kotlyar and M. Moudgill, "Detecting overflow detection," in *Proceedings of the 2nd IEEE/ACM/IFIP International Conference on Hardware/Software Codesign and Systems Synthesis (CODES+ISSS '04)*, pp. 36–41, Stockholm, Sweden, September 2004.

[22] P. I. Balzola, M. Schulte, J. Ruan, J. Glossner, and E. Hokenek, "Design alternatives for parallel saturating multioperand adders," in *Proceedings of IEEE International Conference on Computer Design: VLSI in Computers and Processors (ICCD '01)*, pp. 172–177, Austin, Tex, USA, September 2001.

[23] Y. Lin, H. Lee, M. Woh, et al., "SODA: a low-power architecture for software radio," in *Proceedings of the 33rd International Symposium on Computer Architecture (ISCA '06)*, pp. 89–100, Boston, Mass, USA, June 2006.

[24] A. Lodi, A. Cappelli, M. Bocchi, et al., "XiSystem: a XiRisc-based SoC with reconfigurable IO module," *IEEE Journal of Solid-State Circuits*, vol. 41, no. 1, pp. 85–96, 2006.

[25] A. Duller, G. Panesar, and D. Towner, "Parallel processing - the picoChip way!," in *Communicating Process Architectures (CPA '03)*, pp. 125–138, Enschede, The Netherlands, September 2003.

[26] B. Mohebbi, E. C. Filho, R. Maestre, M. Davies, and F. J. Kurdahi, "A case study of mapping a software-defined radio (SDR) application on a reconfigurable DSP core," in *Proceedings of the 1st IEEE/ACM/IFIP International Conference on Hardware/Software Codesign and System Synthesis*, pp. 103–108, Newport Beach, Calif, USA, October 2003.

[27] T. Ungerer, B. Robič, and J. Šilc, "A survey of processors with explicit multithreading," *ACM Computing Surveys*, vol. 35, no. 1, pp. 29–63, 2003.

[28] B. J. Smith, "The architecture of HEP," in *Parallel MIMD Computation: HEP Supercomputer and Its Applications*, J. S. Kowalik, Ed., pp. 41–55, MIT Press, Cambridge, Mass, USA, 1985.

[29] T. E. Mankovic, V. Popescu, and H. Sullivan, "CHoPP principles of operations," in *Proceedings of the 2nd International Supercomputer Conference*, pp. 2–10, Mannheim, Germany, May 1987.

[30] D. M. Tullsen, S. J. Eggers, and H. M. Levy, "Simultaneous multithreading: maximizing on-chip parallelism," in *Proceedings of the 22nd Annual International Symposium on Computer Architecture (ISCA '95)*, pp. 392–403, Santa Margherita Ligure, Italy, June 1995.

[31] D. Parson, P. Beatty, J. Glossner, and B. Schlieder, "A framework for simulating heterogeneous virtual processors," in *Proceedings of the 32nd Annual Simulation Symposium*, pp. 58–67, San Diego, Calif, USA, April 1999.

[32] R. Leupers, J. Elste, and B. Landwehr, "Generation of interpretive and compiled instruction set simulators," in *Proceedings of the Asia and South Pacific Design Automation Conference (ASP-DAC '99)*, vol. 1, pp. 339–342, Wanchai, Hong Kong, January 1999.

[33] V. Zivojnovic, S. Tjiang, and H. Meyr, "Compiled simulation of programmable DSP architectures," in *Proceedings of the IEEE Workshop on VLSI Signal Processing*, pp. 187–196, Osaka, Japan, October 1995.

[34] S. Pees, A. Hoffmann, V. Zivojnovic, and H. Meyr, "LISA—machine description language for cycle-accurate models of programmable DSP architectures," in *Proceedings of the 36th Annual Design Automation Conference (DAC '99)*, pp. 933–938, New Orleans, La, USA, June 1999.

[35] J. Zhu and D. D. Gajski, "An ultra-fast instruction set simulator," *IEEE Transactions on Very Large Scale Integration (VLSI) Systems*, vol. 10, no. 3, pp. 363–373, 2002.

[36] R. Cmelik and D. Keppel, "Shade: a fast instruction-set simulator for execution profiling," Tech. Rep. UWCSE 93-06-06, University of Washington, Washington, DC, USA, 1993.

[37] A. Nohl, G. Braun, O. Schliebusch, R. Leupers, H. Meyr, and A. Hoffmann, "Design innovations for embedded processors: a universal technique for fast and flexible instruction-set architecture simulation," in *Proceedings of the 39th Design Automation Conference (DAC '02)*, pp. 22–27, ACM Press, New Orleans, La, USA, June 2002.

Hindawi Publishing Corporation
EURASIP Journal on Embedded Systems
Volume 2007, Article ID 86273, 13 pages
doi:10.1155/2007/86273

Research Article

A Shared Memory Module for Asynchronous Arrays of Processors

Michael J. Meeuwsen, Zhiyi Yu, and Bevan M. Baas

Department of Electrical and Computer Engineering, University of California, Davis, CA 95616-5294, USA

Received 1 August 2006; Revised 20 December 2006; Accepted 1 March 2007

Recommended by Gang Qu

A shared memory module connecting multiple independently clocked processors is presented. The memory module itself is independently clocked, supports hardware address generation, mutual exclusion, and multiple addressing modes. The architecture supports independent address generation and data generation/consumption by different processors which increases efficiency and simplifies programming for many embedded and DSP tasks. Simultaneous access by different processors is arbitrated using a least-recently-serviced priority scheme. Simulations show high throughputs over a variety of memory loads. A standard cell implementation shares an 8 K-word SRAM among four processors, and can support a 64 K-word SRAM with no additional changes. It cycles at 555 MHz and occupies 1.2 mm^2 in 0.18 μm CMOS.

1. INTRODUCTION

The memory subsystem is a key element of any computational machine. The memory retains system state, stores data for computation, and holds machine instructions for execution. In many modern systems, memory bandwidth is the primary limiter of system performance, despite complex memory hierarchies and hardware driven prefetch mechanisms.

Coping with the intrinsic gap between processor performance and memory performance has been a focus of research since the beginning of the study of computer architecture [1]. The fundamental problem is the infeasibility of building a memory that is both large and fast. Designers are forced to reduce the sizes of memories for speed, or processors must pay long latencies to access high capacity storage. As memory densities continue to grow, memory performance has improved only slightly; processor performance, on the other hand, has shown exponential improvements over the years. Processor performance has increased by 55 percent each year, while memory performance increases by only 7 percent [2]. The primary solution to the memory gap has been the implementation of multilevel memory hierarchies.

In the embedded and signal processing domains, designers may use existing knowledge of system workloads to optimize the memory system. Typically, these systems have smaller memory requirements than general purpose computing loads, which makes alternative architectures attractive.

This work explores the design of a memory subsystem for a recently introduced class of multiprocessors that are composed of a large number of synchronous processors clocked asynchronously with respect to each other. Because the processors are numerous, they likely have fewer resources per processor, including instruction and data memory. Each processor operates independently without a global address space. To efficiently support applications with large working sets, processors must be provided with higher capacity memory storage. The *Asynchronous Array of simple Processors (AsAP)* [3] is an example of this class of chip multiprocessors.

To maintain design simplicity, scalability, and computational density, a traditional memory hierarchy is avoided. In addition, the low locality in tasks such as those found in many embedded and DSP applications, makes the cache solution unattractive for these workloads. Instead, directly-addressable software-managed memories are explored. This allows the programmer to efficiently manage the memory hierarchy explicitly.

The main requirements for the memory system are the following:

(1) the system must provide high throughput access to high capacity random access memory,

(2) the memory must be accessible from multiple asynchronous clock domains,

(3) the design must easily scale to support arbitrarily large memories, and

(4) the impact on processing elements should be minimized.

The remainder of this work is organized as follows. In Section 2, the current state of the art in memory systems is reviewed. Section 3 provides an overview of an example processor array without shared memories. Section 4 explores the design space for memory modules. Section 5 describes the design of a buffered memory module, which has been implemented using a standard cell flow. Section 6 discusses the performance and power of the design, based on high level synthesis results and simulation. Finally, the paper concludes with Section 7.

2. BACKGROUND

2.1. Memory system architectures

Although researchers have not been able to stop the growth of the processor/memory gap, they have developed a number of architectural alternatives to increase system performance despite the limitations of the available memory. These solutions range from traditional memory hierarchies to intelligent memory systems. Each solution attempts to reduce the impact of poor memory performance by storing the data needed for computation in a way that is easily accessible to the processor.

2.1.1. Traditional memory hierarchies

The primary solution to the processor/memory gap has been to introduce a local cache memory, exploiting spatial and temporal locality evident in most software programs. Caches are small fast memories that provide the processor with a local copy of a small portion of main memory. Caches are managed by hardware to ensure that the processor always sees a consistent view of main memory.

The primary advantage of the traditional cache scheme is ease of programming. Because caches are managed by hardware, programs address a single large address space. Movement of data from main memory to cache is handled by hardware and is transparent to software.

The primary drawback of the cache solution is its high overhead. Cache memories typically occupy a significant portion of chip area and consume considerable power. Cache memories do not add functionality to the system—all storage provided is redundant, and identical data must be stored elsewhere in the system, such as in main memory or on disk.

2.1.2. Alternative memory architectures

Scratch Pad Memories are a cache alternative not uncommonly found in embedded systems [4]. A scratch-pad memory is an on-chip SRAM with a similar size and access time as an L1 (level 1) cache. Scratch pad memories are unlike caches

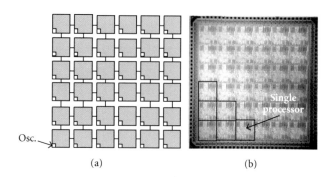

FIGURE 1: Block diagram and chip micrograph of the AsAP chip multiprocessor.

in that they are uniquely mapped to a fixed portion of the system's address space. Scratch-pad memory may be used in parallel with a cache or alone [5]. Banakar et al. report a typical power savings of 40 percent when scratch-pad memories are used instead of caches [4].

Others have explored alternatives to traditional memory hierarchies. These include architectures such as Intelligent RAM (IRAM) [6] and Smart Memories [7].

3. AN EXAMPLE TARGET ARCHITECTURE: AsAP

An example target architecture for this work is a chip multiprocessor called an Asynchronous Array of simple Processors (AsAP) [3, 8, 9]. An AsAP system consists of a two-dimensional array of homogeneous processing elements as shown in Figure 1. Each element is a simple CPU, which contains its own computation resources and executes its own locally stored program. Each processing element has a local clock source and operates asynchronously with respect to the rest of the array. The Globally Asynchronous Locally Synchronous (GALS) [10] nature of the array alleviates the need to distribute a high speed clock across a large chip. The homogeneity of the processing elements makes the system easy to scale as additional tiles can be added to the array with little effort.

Interprocessor communication within the array occurs through dual-clock FIFOs [11] on processor boundaries. These FIFOs provide the required synchronization, as well as data buffers for rate matching between processors. The interconnection of processors is reconfigurable.

Applications are mapped to AsAP by partitioning computation into many small tasks. Each task is statically mapped onto a small number of processing elements. For example, an IEEE 802.11a baseband transmitter has been implemented on a 22-processor array [9], and a JPEG encoder has been implemented on a 9-processor array.

AsAP processors are characterized by their very small memory resources. Small memories minimize power and area while increasing the computational density of the array. No memory hierarchy exists, and memory is managed entirely by software. Additionally, there is no global address space, and all interprocessor communication must occur through the processors' input FIFOs.

Each processor tile contains memory for 64 32-bit instructions and 128 16-bit words. With only 128 words of randomly-accessible storage in each processor, the AsAP architecture is currently limited to applications with small working sets.

4. DESIGN SPACE EXPLORATION

A wide variety of design possibilities exist for adding larger amounts of memory to architectures like AsAP. This section describes the design space and design selection based on estimated performance and flexibility.

In exploring the design space, parameters can be categorized into three roughly orthogonal groups.

(1) *Physical design* parameters, such as memory capacity and module distribution have little impact on the design of the memory module itself, but do determine how the module is integrated into the processing array.
(2) *Processor interface* parameters, such as clock source and buffering have the largest impact on the module design.
(3) *Reconfigurability* parameters allow design complexity to be traded off for additional flexibility.

4.1. Key memory parameters

4.1.1. Capacity

Capacity is the amount of storage included in each memory module. Memory capacity is driven by application requirements as well as area and performance targets. The lower bound on memory capacity is given by the memory requirements of targeted applications while die area and memory performance limit the maximum amount of memory. Higher capacity RAMs occupy more die area, decreasing the total computational density of the array. Larger RAMs also limit the bandwidth of the memory core.

It is desirable to implement the smallest possible memory required for the targeted applications. These requirements, however, may not be available at design time. Furthermore, over-constraining the memory capacity limits the flexibility of the array as new applications emerge. Hence, the scalability of the memory module design is important, allowing the memory size to be chosen late in the design cycle and changed for future designs with little effort.

4.1.2. Density

Memory module density refers to the number of memory modules integrated into an array of a particular size, and is determined by the size of the array, available die area, and application requirements. Typically, the number of memory modules integrated into an array is determined by the space available for such modules; however, application level constraints may also influence this design parameter. Assuming a fixed memory capacity per module, additional modules may be added to meet minimum memory capacity requirements.

Also, some performance increase can be expected by partitioning an application's data among multiple memory modules due to the increased memory bandwidth provided by each module. This approach to increasing performance is not always practical and does not help if the application does not saturate the memory interface. It also requires a high degree of parallelism among data as communication among memory modules may not be practical.

4.1.3. Distribution

The distribution of memory modules within the array can take many forms. In general, two topological approaches can be used. The first approach leaves the processor array intact and adds memory modules in rows or columns as allowed by available area resources. Processors in the array maintain connectivity to their nearest neighbors, as if the memory modules were not present. The second approach replaces processors with memory modules, so that each processor neighboring a memory module loses connectivity to one processor. These strategies are illustrated in Figure 2.

4.1.4. Clock source

Because the targeted arrays are GALS systems, the clock source for the memory module becomes a key design parameter. In general, three distinct possibilities exist. First, the memory module can derive its clock from the clock of a particular processor. The memory would then be synchronous with respect to this processor. Second, the memory can generate its own unique clock. The memory would be asynchronous to all processors in the array. Finally, the memory could be completely asynchronous, so that no clock would be required. This solution severely limits the implementation of the memory module, as most RAMs provided in standard cell libraries are synchronous.

4.1.5. Address source

The address source for a memory module has a large impact on application mapping and performance. To meet the random access requirement, processors must be allowed to supply arbitrary addresses to memory. (1) The obvious solution uses the processor producing or consuming the memory data as the address source. The small size of the targeted processors, however, makes another solution attractive. (2) The address and data streams for a memory access can also be partitioned among multiple processors. A single processor can potentially be used to provide memory addresses, while other processors act as data sources and data sinks. This scheme provides a potential performance increase for applications with complex addressing needs because the data processing and address generation can occur in parallel. (3) A third possible address source is hardware address generators, which typically speed up memory accesses significantly, but must be built into hardware. To avoid unnecessary use of power and die area, only the most commonly used access patterns should be included in hardware.

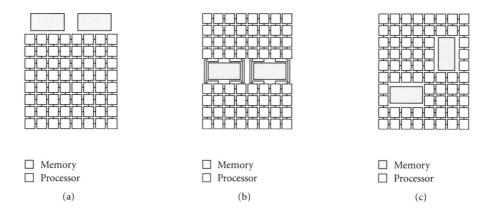

FIGURE 2: Various topologies for distribution of memories in a processor array. Processor connectivity is maintained when (a) memories are added to the edge of the array, or (b) the array is split to make room for a row of memories. Processor connectivity is lost when (c) processor tiles are replaced by memory tiles.

4.1.6. Buffering

The implementation of buffers for accesses to the memory module provides another design parameter. Buffers may be used between a processor and a memory module for latency hiding, synchronization, or rate matching. Without some level of buffering, processors are tightly coupled to the memory interface, and prefetching of data is difficult.

4.1.7. Sharing

The potentially large number of processors in a processing array makes the sharing of memories among processors attractive. In this context, shared memory serves two distinct purposes. First, as in more traditional computing, shared memory can serve as a communication medium among simultaneous program threads. Also, in our context, sharing a memory among multiple processors can enable higher utilization of available memory bandwidth in cases where a single thread is unable to saturate the memory bus. In either case, synchronization mechanisms are required to guarantee mutual exclusion when memory is shared.

4.1.8. Inter-parameter dependencies

There are strong dependencies among the four parameters described in the preceding four subsections (clock source, address source, buffering, and sharing). Selecting a value for one of the parameters limits the feasible values of the other parameters. This results in the existence of two distinct archetype designs for the processor interface. Other design options tend to be hybrids of these two models and often have features that limit usefulness.

Type I: bufferless memory

The first design can be derived by forcing a *bufferless* implementation. Without buffers, there is no way to synchronize across clock boundaries, so the memory module must be synchronous to the interfacing processor. Because processors are asynchronous to one another, sharing the memory is no longer feasible, and using an alternate processor as an address source is not possible. The resulting design is a memory module that couples tightly to a single processor. Because there is no buffering, memory accesses are either tightly integrated into the processor's pipeline or carefully timed to avoid overwriting data.

Type II: buffered memory

The second design is, in some respects, the dual of the first. We can arrive at this design by requiring that the memories be *shareable*. Because processors exist in different clock domains, dual-clock FIFOs must be used to synchronize across clock boundaries. To avoid tying the memory clock speed to an arbitrary processor (which would defeat the fundamental purpose of GALS clocking—namely, to allow independent frequency adjustment of blocks), the memory module should supply its own clock. An independent processor could easily be used as an address source with the appropriate hardware in place. This design effectively isolates the memory module from the rest of the array, has few dependencies on the implementation of the processors, and does not impact the performance of any processors not accessing the memory.

4.2. Degree of configurability

The degree of configurability included in the memory-processor interconnect, as well as in the memory module itself can be varied independently of the memory module design. To some degree, the level of configurability required in the interconnect is a function of the number of processors in the array, and their distances from the memory module. For small arrays, hardwired connections to the memory module may make sense. For large arrays with relatively few memory modules, additional configurability is desirable to avoid limiting the system's flexibility.

The configurability of the memory module itself allows trade offs in performance, power, and area for flexibility. Examples of configurability at the module level cover a broad range and are specific to the module's design. Some examples of configurable parameters are the address source used for memory accesses and the direction of synchronization FIFOs in a locally clocked design.

4.3. Design selection

The remainder of this work describes a buffered memory solution. This design was chosen based on the flexibility in addressing modes and the ability to share the memory among multiple processors. These provide a potential performance increase by allowing redistribution of the address generation workload, and by exploiting parallelism across large datasets. The relative area overhead impact of the additional logic can be reduced if the RAM core used in the memory module has a high capacity and thus the FIFO buffers become a small fraction of the total module area. The performance impact of additional memory latency can potentially be reduced or eliminated by appropriate software task partitioning or techniques such as data prefetching.

5. FIFO-BUFFERED MEMORY DESIGN

This section describes the design and implementation of a FIFO-buffered memory module suitable for sharing among independently-clocked interfaces (typically processors). The memory module has its own local clock source, and communicates with external blocks via dual clock FIFOs. As described in Section 4.3, this design was selected based on its flexibility in addressing modes and the potential speedup for applications with a high degree of parallelism across large datasets.

5.1. Overview

The prototype described in this section allows up to four external blocks to access the RAM array. The design supports a memory size up to 64 K 16-bit words with no additional modifications.

Processors access the memory module via input ports and output ports. *Input ports* encapsulate the required logic to process incoming requests and utilize a dual-clock FIFO to reliably cross clock domains. Each input port can assume different modes, changing the method of memory access. The memory module returns data to the external block via an *output port*, which also interfaces via a dual-clock FIFO.

A number of additional features are integrated into the memory module to increase usability. These include multiple port modes, address generators, and mutual exclusion (mutex) primitives. A block diagram of the FIFO-buffered memory is shown in Figure 3. This diagram shows the high-level interaction of the input and output ports, address generators, mutexes, and SRAM core. The theory of operation for this module is described in Section 5.2. The programming interface to the memory module is described in Section 5.3.

5.2. Theory of operation

The operation of the FIFO-buffered memory module is based on the execution of requests. External blocks issue requests to the memory module by writing 16-bit command tokens to the input port. The requests instruct the memory module to carry out particular tasks, such as memory writes or port configuration. Additional information on the types of requests and their formats is provided in Section 5.3. Incoming requests are buffered in a FIFO queue until they can be issued. While requests issued into a single port execute in FIFO order, requests from multiple processors are issued concurrently. Arbitration among conflicting requests occurs before allowing requests to execute.

In general, the execution of a request occurs as follows. When a request reaches the head of its queue it is decoded and its data dependencies are checked. Each request type has a different set of requirements. A memory read request, for example, requires adequate room in the destination port's FIFO for the result of the read; a memory write, on the other hand, must wait until valid data is available for writing. When all such dependencies are satisfied, the request is issued. If the request requires exclusive access to a shared resource, it requests access to the resource and waits for acknowledgment prior to execution. The request blocks until access to the resource is granted. If the request does not access any shared resources, it executes in the cycle after issue. Each port can potentially issue one request per cycle, assuming that requests are available and their requirements are met.

The implemented memory module supports all three address sources detailed in Section 4.1.5. These are (1) one processor providing addresses and data, (2) two processors with one providing addresses and the other handling data, and (3) hardware address generators. All three support bursts of 255 memory reads or writes with a single request. These three modes provide high efficiency in implementing common access patterns without preventing less common patterns from being used.

Because the memory resources of the FIFO-buffered memory are typically shared among multiple processors, the need for interprocess synchronization is anticipated. To this end, the memory module includes four mutex primitives in hardware. Each mutex implements an atomic single-bit test and set operation, allowing easy implementation of simple locks. More complex mutual exclusion constructs may be built on top of these primitives using the module's memory resources.

5.3. Processor interface

External blocks communicate with the memory module via dedicated memory ports. Each of these ports may be configured to connect to one input FIFO and one output FIFO in the memory module. These connections are independent, and which of the connections are established depends on the size of the processor array, the degree of reconfigurability implemented, and the specific application being mapped.

An external block accesses the memory module by writing 16-bit words to one of the memory module's input

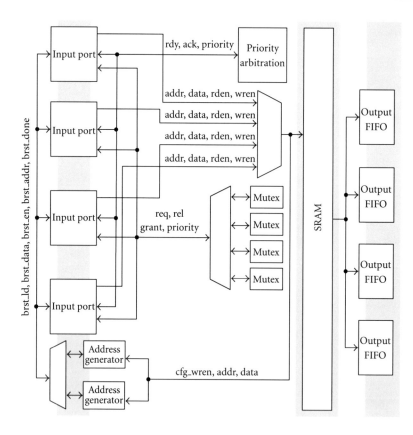

FIGURE 3: FIFO-buffered memory block diagram. Arrows show the direction of signal flow for the major blocks in the design. Multiplexers allow control of various resources to be switched among input ports. The gray bars approximate the pipeline stages in the design.

FIFOs. In general, these words are called *tokens*. One or more tokens make up a *request*. A request instructs the memory module to perform an action and consists of a command token, and possibly one or more data tokens. The requests issued by a particular processor are always executed in FIFO order. Concurrent requests from multiple processors may be executed in any order. If a request results in data being read from memory, this data is written to the appropriate output FIFO where it can be accessed by the appropriate block.

5.3.1. Request types

The FIFO-buffered memory supports eight different request types. Each request type utilizes different resources within the memory module. In addition, some requests are blocking, meaning that they must wait for certain conditions to be satisfied before they complete. To maintain FIFO ordering of requests, subsequent requests cannot proceed until a blocking request completes.

(1)-(2) Memory read and write requests cause a single word memory access. The request blocks until the access is completed. (3)-(4) Configuration requests enable setup of module ports and address generators. (5)-(6) Burst read and write requests are used to issue up to 255 contiguous memory operations using an address generator. (7)-(8) Mutex request and release commands are used to control exclusive use of a mutual exclusion primitive—which can be used for

synchronization among input ports or in the implementation of more complex mutual exclusion constructs.

5.3.2. Input port modes

Each input port in the FIFO-buffered memory module can operate in one of three modes. These modes affect how incoming memory and burst requests are serviced. Mode information is set in the port configuration registers using a port configuration request. These registers are unique to each input port, and can only be accessed by the port that contains them.

Address-data mode is the most fundamental input port mode. In this mode, an input port performs memory reads and writes independently. The destination for memory reads is programmable, and is typically chosen so that the output port and input port connect to the same external block, but this is not strictly required.

A memory write is performed by first issuing a memory write request containing the write address. This request must be immediately followed by a data token containing the data to be written to memory. In the case of a burst write, the burst request must be immediately followed by the appropriate number of data tokens. Figure 4(a) illustrates how writes occur in address-data mode.

A memory read is performed by first issuing a memory read request, which contains the read address. The value read

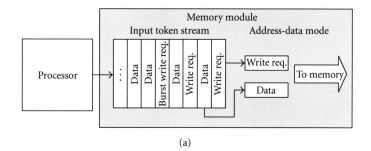

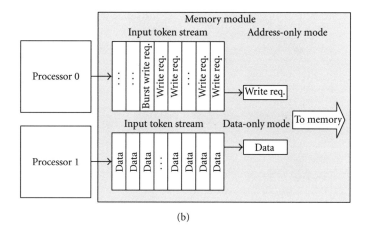

FIGURE 4: Memory writes in (a) address-data and (b) address-only mode. (a) In *address-data* mode, each port provides both addresses and data. Memory writes occur independently, and access to the memory is time-multiplexed. Two tokens must be read from the same input stream to complete a write. (b) In *address-only* mode, write addresses are supplied by one port, and data are supplied by another. Memory writes are coupled, so there is no need to time-multiplex the memory among ports. One token must be read from each input stream to complete a write.

from memory is then written to the specified output FIFO. The same destination is used for burst reads.

In *address-only* mode, an input port is paired with an input port in *data-only* mode to perform memory writes. This allows the tasks of address generation and data generation to be partitioned onto separate external blocks.

In address-only mode, a memory write is performed by issuing a memory write request containing the write address. In contrast to operation in address-data mode, however, this request is not followed by a data token. Instead, the next valid data token from the input port specified by a programmable configuration register is written to memory. Synchronization between input ports is accomplished by maintaining FIFO order of incoming tokens. It is the programmer's responsibility to ensure that there is a one to one correspondence between write requests in the address-only port and data tokens in the data-only port. Figure 4(b) illustrates how writes occur.

An input port in data-only mode acts as a slave to the address-only input port to which it provides data. All request types, with the exception of port configuration requests, are ignored when the input port is in data-only mode. Instead all incoming tokens are treated as data tokens. The programmer must ensure that at any one time, at most one input port is configured to use a data-only port as a data source.

As previously mentioned, the presented memory module design directly supports memory arrays up to 64 K words. This is due solely to a 16-bit interface in the AsAP processor, and therefore a 16-bit memory address in a straightforward implementation. The supported address space can clearly be increased by techniques such as widening the interface bus or implementing a paging scheme.

Another dimension of memory module scaling is to consider connecting more than four processors to a module. This type of scaling begins to incur significant performance penalties (in tasks such as port arbitration) as the number of ports scales much beyond four. Instead, the presented memory module is much more amenable to replication throughout an array of processors—providing high throughput to a small number of local processors while presenting no barriers to the joining of multiple memories through software or interconnect reconfiguration, albeit with a potential increase in programming complexity depending on the specific application.

6. IMPLEMENTATION RESULTS

The FIFO-buffered memory module described in Section 5 has been described in Verilog and synthesized with a 0.18 μm CMOS standard cell library. A standard cell implementation

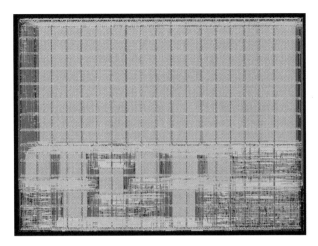

FIGURE 5: Layout of a 8192-word × 16-bit 0.18 μm CMOS standard cell FIFO-buffered memory module implementation. The large SRAM is at the top of the layout and the eight 32-word FIFO memories are visible in the lower region.

has been completed and is shown in Figure 5. The design is fully functional in simulation.

Speed, power, and area results were estimated from high-level synthesis. In addition, the design's performance was analyzed with RTL level simulation. This section discusses these results.

6.1. Performance results

System performance was the primary metric motivating the memory module design. Two types of performance are considered. First, the system's peak performance, as dictated by the maximum clock frequency, peak throughput, and latency is calculated. A more meaningful result, however, is the performance of actual programs accessing the memory. Both of these metrics are discussed in the following subsections.

6.1.1. Peak performance

The peak performance of the memory module is a function of the maximum clock frequency, and the theoretical throughput of the design. The FIFO-buffered memory module is capable of issuing one memory access every cycle, assuming that requests are available and their data dependencies are met. In address-data mode, memory writes require a minimum of two cycles to issue, but this penalty can be avoided by using address generators or the address-only port mode, or by interleaving memory requests from multiple ports. If adequate processing resources exist to supply the memory module with requests, the peak memory throughput is one word access per cycle. Synthesis results report a maximum clock frequency of 555 MHz. At this clock speed, the memory's peak throughput is 8.8 Gbps with 16-bit words.

The worst case memory latency is for the memory read request. There are contributions to this latency in each of the system's clock domains. In the external block's clock domain, the latency includes one FIFO write latency, one FIFO read latency, and the additional latency introduced by the memory port. In the memory's clock domain, the latency includes one FIFO read latency, the memory module latency, and one FIFO write latency.

The minimum latency of the memory module is given by the number of pipe stages between the input and output ports. The presented implementation has four pipe stages. The number of stages may be increased to add address decoding stages for larger memories.

The latency of FIFO reads and writes is dependent on the number of pipe stages used to synchronize data across the clock boundary between the read side and the write side. In AsAP's FIFO design, the number of synchronization stages is configurable at runtime. When a typical value of three stages is used, the total FIFO latency is four cycles per side. When the minimum number of stages is used, the latency is reduced to three cycles per side. A latency of four cycles is assumed in this work.

The latency of the memory port depends on the number of stages introduced between the processor and the memory to account for wire delays. The minimum latency of the memory port is two cycles. This latency could be decreased by integrating the memory port more tightly with the processor core datapath. This approach hinders the use of a prefetch buffer to manage an arbitrary latency from processor to memory, and is only practical if the latency can be constrained to a single cycle.

Summing the latency contributions from each clock domain, the total latency of a memory read is,

$$
\begin{aligned}
L_{\text{proc}} &= L_{\text{FIFO-wr}} + L_{\text{FIFO-rd}} + L_{\text{mem-port}}, \\
L_{\text{mem}} &= L_{\text{FIFO-rd}} + L_{\text{mem-module}} + L_{\text{FIFO-wr}}, \\
L_{\text{total}} &= L_{\text{mem}} + L_{\text{proc}}.
\end{aligned}
\tag{1}
$$

For the presented design and typical configurations, the latency is 10 processor cycles and 13 memory cycles. If the blocks are clocked at the same frequency, this is a minimum latency of 23 cycles. Additional latency may be introduced by processor stalls, memory access conflicts, or data dependencies. The latency is slightly higher than typical L2 cache latencies, which are on the order of 15 cycles [2], due to the communication overhead introduced by the FIFOs. This high latency can be overcome by issuing multiple requests in a single block. Because requests are pipelined, the latency penalty occurs only once per block.

6.1.2. Actual performance

To better characterize the design's performance, the memory module was exercised with two generic and variable workloads: a *single-element workload* and a *block workload*. The number of instructions in both test kernels is varied to simulate the effect of varying computation loads for each application. Figure 6 gives pseudocode for the two workloads.

The single-element workload performs a copy of a 1024-element array and contains three phases. First, a burst write

```
Write initial array    ⎡ for i = 0 : 1023
                       │ mem_wr(a + i)   // wr command
                       │ wr_data = input // wr data
                       ⎣ next i
                         for i = 0 : 1023
                         mem_rd(a + i)   // rd command
                         mem_wr(b + i)   // wr command

# NOPs models          ⎡ NOP
additional             │ · · ·
computation load       ⎣ NOP
                         temp = rd_data   // rd data
                         wr_data = temp   // wr data
                         next i

Read result            ⎡ for i = 0 : 1023
                       │ mem_rd(b + i)    // rd command
                       │ output = rd_data // rd data
                       ⎣ next i
```

(a) Single element workload

```
                                       for i = 0 : 1023
# NOPs models          ⎡ NOP
additional address     │ · · ·
computation load       ⎣ NOP

                       mem_wr(a + i)  // wr command

# NOPs models          ⎡ NOP
additional data        │ · · ·
computation load       ⎣ NOP

                       wr_data = input  // wr data
                       next i

# NOPs models          ⎡ NOP
additional address     │ · · ·
computation load       ⎣ NOP

                       mem_rd(a + i)  // rd command

# NOPs models          ⎡ NOP
additional data        │ · · ·
computation load       ⎣ NOP

                       output = rd_data  // rd data
                       next i
```

(b) Block workload

FIGURE 6: Two workloads executed on external processors are used for performance characterization. Pseudo-code for the two workloads is shown for processors in address-data mode. In each workload, the computational load per memory transaction is simulated and varied by adjusting the number of NOPs in the main kernel. The block workload is also tested in address-only/data-only mode (not shown here) where the code that generates memory requests, and the code that reads and writes data is partitioned appropriately. *mem_rd()* and *mem_wr()* are read and write commands being issued with the specified address. *rd_data* reads data from the processor's memory port, and *wr_data* writes data to the processor's memory port.

is used to load the source array into the processor. Second, the array is copied element by element, moving one element per loop iteration. Finally, the resulting array is read out with a burst read. The number of instructions in the copy kernel is varied to simulate various computational loads. The single-element kernel is very sensitive to memory read latency because each memory read must complete before another can be issued. To better test throughput rather than latency, the block test is used. This workload first writes 1024 memory words, and then reads them back.

In addition, three coding approaches are compared. The first uses a single processor executing a single read or write per loop iteration. The second uses burst requests to perform memory accesses. The third approach partitions the task among two processors in address-only and data-only modes. One processor issues request addresses, while the other manages data flow. Again, the number of instructions in each kernel is varied to simulate various computational loads.

Figure 7 shows the performance results for the single-element workload running on a single processor at different clock speeds. For small workloads, the performance is dominated by the memory latency. This occurs because each iteration of the loop must wait for a memory read to complete before continuing. A more efficient coding of the kernel could overcome this latency using loop unrolling techniques. This may not always be practical, however, due to limited code and data storage. The bend in each curve occurs at the location where the memory latency is matched to the computational workload. Beyond this point, the performance scales with the complexity of computation. The processor's clock speed has the expected effect on performance. At high frequencies, the performance is still limited by memory latency, but larger workloads are required before the computation time overcomes the read latency. The latency decreases slightly at higher processor frequencies because the component of the latency in the processor's clock domain is reduced. The slope of the high-workload portion of the curve is reduced because the relative impact of each additional instruction is less at higher frequencies.

For highly parallel workloads, the easiest way to improve performance is to distribute the task among multiple processors. Figure 8 shows the result of distributing the single-element workload across one, two, and four processors. In this case, the 1024 copy operations are divided evenly among all of the processors. When mapped across multiple processors, one processor performs the initial array write, and one processor performs the final array read. The remainder of the computation is distributed uniformly among the processors. Mutexes are used to ensure synchronization

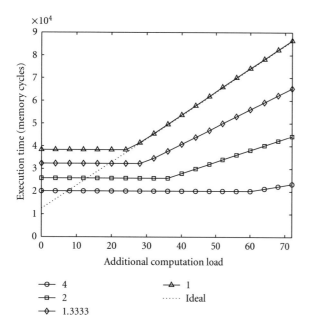

FIGURE 7: Effect of computational load and clock speed on performance. The figure shows the execution time of the single-element workload for a single processor clocked at 1, 1.33, 2, and 4 times the memory speed. The dotted line represents the theoretical maximum performance for the workload operating on a single processor clocked at the same speed as the memory.

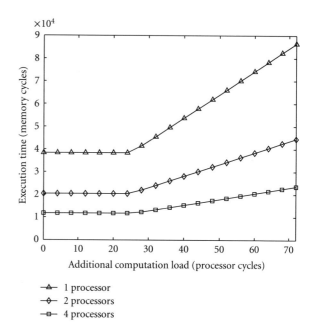

FIGURE 8: Effect of number of processors on performance. The figure shows the execution time of the single-element workload for 1, 2, and 4 processors clocked at the same frequency as the memory. The execution time for each case includes some fixed overhead to initialize and read the source and destination arrays. Multiple processor cases have additional overhead for synchronization among processors.

between the initialization, copy, and read-out phases of execution.

When the single-element workload is shared among processors, the application's performance is increased at the cost of additional area and power consumed by the additional processors. For small computation loads, the effective read latency is reduced. Although each read still has the same latency, the reads from each processor are issued concurrently. Hence, the total latency suffered scales inversely with the number of processors used. For loads where latency is dominated by computation cost, the impact of the computation is reduced, because multiple iterations of the application kernel run concurrently on the various processors. Note that the point where computation load begins to dominate latency is constant, regardless of the number of processors used. The relative latency depends only on the relative clock speeds of the processors and memories, and not on the distribution of computation.

Figure 9 shows the performance of the three addressing schemes for the block workload when the processors and memory are clocked at the same frequency. For small workloads, the address-data mode solution is dominated by read latency and write workload. Because writes are unaffected by latency, the computation load has an immediate effect. For large workloads, the execution time is dominated by the computation load of both reads and writes. To illustrate the appropriateness of the synthetic workloads, three key algorithms (1024-tap FIR filter, 512-point complex FFT, and a

viterbi decoder) are modeled and shown on the plot. While these applications are not required to be written conforming to the synthetic workloads, the versions shown here are very reasonable implementations.

The address generator and address-only/data-only solutions decouple the generation of memory read requests from the receipt of read data. This allows requests to be issued far in advance, so the read latency has little effect. There is also a slight performance increase because the number of instructions in each kernel is reduced.

The address generator solution outperforms the single cycle approach, and does not require the allocation of additional processors. This is the preferred solution for block accesses that can be mapped to the address generation hardware. For access patterns not supported by the address generators, similar performance can be obtained by generating the addresses with a processor in address-only mode. This requires the allocation of an additional processor, which does incur an additional cost.

Address only mode allows arbitrary address generation capability at the cost of an additional processor. This method eases implementation of latency-insensitive burst reads without requiring partitioning of the data computation. This method is limited by the balance of the address and data computation loads. If the address and data processors run at the same speed, whichever task carries the highest computation load dominates the system performance. This can be seen in Figure 10.

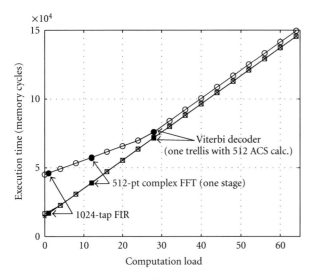

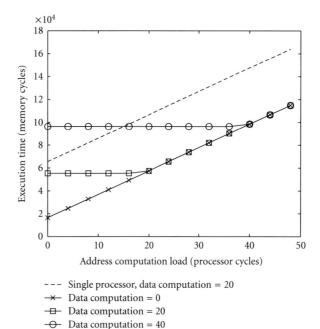

FIGURE 9: Effect of address mode on block performance. The figure shows the execution time of the block workload for a single processor in address-data mode, a single processor utilizing address generator hardware, and two processors, one in address-only mode and one in data-only mode. Both the address generator and address-only mode solutions outperform the address-data mode solution if the work load is dominated by the memory latency. Note that the address generator and address-only performances are roughly equal. Three real applications are shown to validate the synthetic workloads.

FIGURE 10: Effect of address load on address-only mode performance. The figure shows the execution time of the block workload for a single processor in address-data mode, and two processors, one in address-only mode and one in data-only mode. The address calculation workload is varied for each case. Each curve represents a fixed data computation workload. The memory module and processors share the same clock frequency.

Partitioning an application among multiple processors in address-data mode typically outperforms a mapping using the same number of processors in address-only or data-only mode. This occurs because the number of iterations of the application kernel required per processor is reduced. This reduces the application's sensitivity to computation loads. Address-only mode is most useful when address computation and data computation are of similar complexities, when code space limitations prevent the two tasks from sharing the same processor, or when the application lacks adequate data parallelism to distribute the computation otherwise.

Figure 11 compares the performance of the block workload when computation is distributed across two processors. A mapping with two address-data mode processors outperforms address-only and data-only partitioning in most cases. If address and data computation loads are mismatched, the greater load dominates the execution time for the address-only/data-only mapping. When the address and data computation loads are similar, the performance gap for the address-only mode mapping is small. Furthermore, for very small computation loads, the address-only mode mapping outperforms the address-data mode mapping because each loop iteration contains fewer instructions.

6.2. Area and power tradeoffs

As with most digital IC designs, area and power are closely related to performance. Generally, performance can be increased at the expense of area and power by using faster devices or by adding parallel hardware. Although the performance of the FIFO-buffered memory module was the first design priority, the power and area results are acceptable. The results discussed are for high-level synthesis of the Verilog model. Some increase is expected during back-end flows. Dynamic power consumption was estimated using activity factors captured during RTL simulation.

6.2.1. Area results

The results of the synthesis flow provide a reasonable estimate of the design's area. The design contains 9713 cells, including hard macros for the SRAM core and FIFO memories. With an 8 K-word SRAM, the cell area of the synthesized design is 1.28 mm². This is roughly equivalent to two and a half AsAP processors. The area after back-end placement and routing is 1.65 mm².

The area of the FIFO buffered memory module is dominated by the SRAM core, which occupies 68.2% of the module's cell area. This implies a 32.8% area overhead to implement the FIFO-buffered design, rather than a simpler SRAM

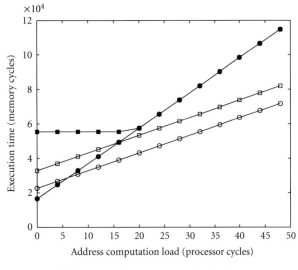

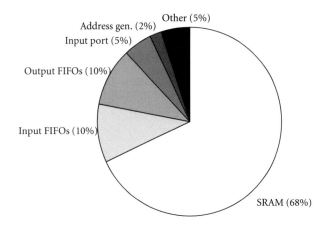

FIGURE 12: Area distribution among submodules. The relative cell area of each group of submodules is shown. The SRAM used in this design is 8 K words. The SRAM consumes 68% of the area; the four input FIFOs occupy 10%; the four output FIFOs occupy 10%. The "other" category includes modules not shown such as configuration, mutex, arbiter, and clock oscillator.

FIGURE 11: Equal area comparison of address modes. The figure shows the execution time of the block workload for two parallel processors in address-data mode, and two processors, one in address-only mode and one in data-only mode. The address calculation workload is varied for each case. Each curve represents a fixed data computation workload. The memory module and processors share the same clock frequency.

interface. This overhead is similar to that reported by Mai et al. for a Smart Memories system with the same memory capacity [12]. The distribution of area among the major blocks of the design is shown in Figure 12. The presented implementation has an SRAM of 8 K words, but the synthesized source design, however, is capable of addressing up to 64 K words. Conservatively assuming the memory size scales linearly, a 64 K-word memory would occupy 94.5% of the module's area. This implies an overhead of only 5.5%, which is easily justified by the performance increase provided by the shared memory.

6.2.2. Power results

In general, accurate power estimation is difficult without physical design details. A reasonable estimate of the design's power consumption can be taken from high level synthesis results and library information. The main limitation of power estimation at this level is obtaining accurate switching activity for the nodes in the design. Switching activity was recorded for the execution of a four processor application that computes $c_j = a_j + 2b_j$ for 1024 points. This application exercises most of the design's functionality.

Power Compiler reports the relative power of each submodule as shown in Figure 13. Absolute power estimates from the tool are believed to be less accurate so we present the relative numbers only. Of the total power consumption,

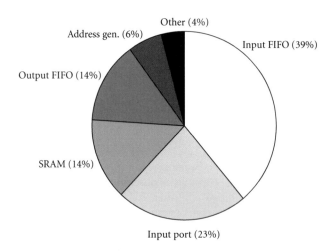

FIGURE 13: Relative power consumption neglecting clock power of submodules. The power is dominated by the input FIFOs (39%) and input ports (23%) as these are the most active blocks in the design. The dynamic power of the SRAM cell is relatively low, but matches well with the datasheet value.

57.1 nW is attributed to cell leakage power. The breakdown for leakage power is shown in Figure 14.

7. CONCLUSION

The design of an asynchronously sharable FIFO-buffered memory module has been described. The module allows a high capacity SRAM to be shared among independently-clocked blocks (such as processors). The memory module shares its memory resources with up to four blocks/processors. This allows the memory to be used for interprocess communication or to increase application performance by parallelizing computation. The addition of addressing modes

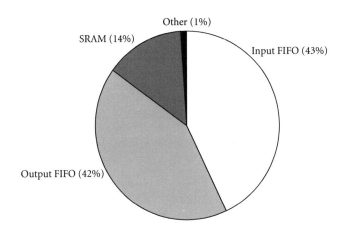

FIGURE 14: Relative leakage power of submodules. As expected, the leakage power is dominated by the memory cells. The eight FIFOs (85%) and the SRAM core (14%) consume nearly all of the module's leakage power.

and hardware address generators increases the system's flexibility when mapping many applications. The FIFO-buffered memory module was described in Verilog, and synthesized with a 0.18 μm CMOS standard cell library. A design with an 8 K-word SRAM has a maximum operating frequency of 555 MHz, and occupies 1.2 mm^2 based on high-level synthesis results. The memory module can service one memory access each cycle, leading to a peak memory bandwidth of 8.8 Gbps.

ACKNOWLEDGMENTS

The authors thank E. Work, T. Mohsenin, R. Krishnamurthy, M. Anders, S. Mathew, and other VCL members; and gratefully acknowledge support from Intel, UC MICRO, NSF Grant no. 0430090, and a UCD Faculty Research Grant.

REFERENCES

[1] A. W. Burks, H. H. Goldstine, and J. von Neumann, "Preliminary discussion of the logical design of an electronic computing instrument," in *Collected Works of John von Neumann*, A. H. Taub, Ed., vol. 5, pp. 34–79, The Macmillan, New York, NY, USA, 1963.

[2] J. L. Hennessy and D. A. Patterson, *Computer Architecture, A Quantitative Approach*, chapter Memory Hierarchy Design, Morgan Kaufmann, San Francisco, Calif, USA, 3rd edition, 2003.

[3] Z. Yu, M. J. Meeuwsen, R. Apperson, et al., "An asynchronous array of simple processors for DSP applications," in *IEEE International Solid-State Circuits Conference (ISSCC '06)*, pp. 428–429, San Francisco, Calif, USA, February 2006.

[4] R. Banakar, S. Steinke, B.-S. Lee, M. Balakrishnan, and P. Marwedel, "Scratchpad memory: a design alternative for cache on-chip memory inembedded systems," in *Proceedings of the 10th International Symposium on Hardware/Software Codesign (CODES '02)*, pp. 73–78, Estes Park, Colo, USA, May 2002.

[5] P. R. Panda, N. D. Dutt, and A. Nicolau, "On-chip vs. off-chip memory: the data partitioning problem in embedded processor-based systems," *ACM Transactions on Design Automation of Electronic Systems*, vol. 5, no. 3, pp. 682–704, 2000.

[6] D. Patterson, T. Anderson, N. Cardwell, et al., "A case for intelligent RAM," *IEEE Micro*, vol. 17, no. 2, pp. 34–44, 1997.

[7] K. Mai, T. Paaske, N. Jayasena, R. Ho, W. J. Dally, and M. A. Horowitz, "Smart memories: a modular reconfigurable architecture," in *Proceedings of the 27th Annual International Symposium on Computer Architecture (ISCA '00)*, pp. 161–171, Vancouver, BC, Canada, June 2000.

[8] B. M. Baas, "A parallel programmable energy-efficient architecture for computationally-intensive DSP systems," in *Proceedings of the 37th Asilomar Conference on Signals, Systems and Computers (ACSSC '03)*, vol. 2, pp. 2185–2192, Pacific Grove, Calif, USA, November 2003.

[9] M. J. Meeuwsen, O. Sattari, and B. M. Baas, "A full-rate software implementation of an IEEE 802.11a compliant digital baseband transmitter," in *Proceedings of IEEE Workshop on Signal Processing Systems (SIPS '04)*, pp. 124–129, Austin, Tex, USA, October 2004.

[10] D. M. Chapiro, *Globally-asynchronous locally-synchronous systems*, Ph.D. thesis, Stanford University, Stanford, Calif, USA, October 1994.

[11] R. W. Apperson, "A dual-clock FIFO for the reliable transfer of high-throughput data between unrelated clock domains," M.S. thesis, University of California, Davis, Davis, Calif, USA, 2004.

[12] K. Mai, R. Ho, E. Alon, et al., "Architecture and circuit techniques for a 1.1-GHz 16-kb reconfigurable memory in 0.18-μm CMOS," *IEEE Journal of Solid-State Circuits*, vol. 40, no. 1, pp. 261–275, 2005.

Hindawi Publishing Corporation
EURASIP Journal on Embedded Systems
Volume 2007, Article ID 85029, 15 pages
doi:10.1155/2007/85029

Research Article

Implementing a WLAN Video Terminal Using UML and Fully Automated Design Flow

Petri Kukkala,[1] Mikko Setälä,[2] Tero Arpinen,[2] Erno Salminen,[2] Marko Hännikäinen,[2] and Timo D. Hämäläinen[2]

[1] *Nokia Technology Platforms, Visiokatu 6, 33720 Tampere, Finland*
[2] *Institute of Digital and Computer Systems, Tampere University of Technology, Korkeakoulunkatu 1, 33720 Tampere, Finland*

Received 28 July 2006; Revised 12 December 2006; Accepted 10 January 2007

Recommended by Gang Qu

This case study presents UML-based design and implementation of a wireless video terminal on a multiprocessor system-on-chip (SoC). The terminal comprises video encoder and WLAN communications subsystems. In this paper, we present the UML models used in designing the functionality of the subsystems as well as the architecture of the terminal hardware. We use the Koski design flow and tools for fully automated implementation of the terminal on FPGA. Measurements were performed to evaluate the performance of the FPGA implementation. Currently, fully software encoder achieves the frame rate of 3.0 fps with three 50 MHz processors, which is one half of a reference C implementation. Thus, using UML and design automation reduces the performance, but we argue that this is highly accepted as we gain significant improvement in design efficiency and flexibility. The experiments with the UML-based design flow proved its suitability and competence in designing complex embedded multimedia terminals.

1. INTRODUCTION

Modern embedded systems have an increasing complexity as they introduce various multimedia and communication functionalities. Novel design methods enable efficient system design with rapid path to prototyping for feasibility analysis and performance evaluation, and final implementation.

High-abstraction level design languages have been introduced as a solution for the problem. Unified modeling language (UML) is converging to a general design language that can be understood by system designers as well as softare and hardware engineers [1]. UML is encouraging the development of model-based design methodologies, such as model driven architecture (MDA) [2, 3] that aims at "portability, interoperability, and reusability through architectural separation of concerns" as stated in [4].

Refining the high-abstraction level models towards a physical implementation requires design automation tools due to the vast design space. This means high investments and research effort in tool development to fully exploit new modeling methodologies. High degree of design automation also requires flexible hardware and software platforms to support automated synthesis and configuration. Hence, versatile hardware/software libraries and run-time environments are needed.

Configurability usually complicates the library development and induces various overheads (execution time, memory usage) compared to manually optimized application-specific solutions. However, we argue that automation is needed to handle the complexity and to allow fast time-to-market, and we have to pay the price. Naturally, the trade-off between high performance and fast development time must be defined case by case.

To meet these design challenges in practice, we have to *define a practical design methodology* for the domain of embedded real-time systems. To *exploit the design methodology*, we have to map the concepts of the methodology to the constructs of a high-abstraction level language. Further, we have to develop design tools and platforms (or adapt existing ones) that *support the methodology* and language.

In this paper, we present an extensive case study for *the implementation of a wireless video terminal using a UML 2.0-based design methodology and fully automated design flow.* The paper introduces UML modeling, tools, and platforms to implement a whole complex embedded terminal with several subsystems. This is a novel approach to exploit UML in the implementation of such a complex design.

The implemented terminal comprises video encoder and wireless local area network (WLAN) communications subsystems, which are modeled in UML. Also, the hardware architecture and the distributed execution of application are modeled in UML. Using these models and *Koski design flow* [5] the terminal is implemented as a multiprocessor system-on-chip (SoC) on a single FPGA.

The paper is organized as follows. Section 2 presents the related work. The Koski design flow is presented in Section 3. Section 4 presents the utilized hardware and software platforms. The wireless video terminal and related UML models are presented in Section 5. The implementation details and performance measurements are presented in Section 6. Finally, Section 7 concludes the paper.

2. RELATED WORK

Since *object management group* (OMG) adopted the UML standard in 1997, it has been widely used in the software industry. Currently, the latest adopted release is known as *UML 2.0* [6]. A number of extension proposals (called *proiles*) have been presented for the domain of real-time and embedded systems design.

The implementation of the wireless video terminal is carried out using the UML-based Koski design flow [5]. UML is used to design both the functionality of the subsystems and the underlying hardware architecture. UML 2.0 was chosen as a design language based on three main reasons. First, previous experiences have shown that UML suits well the implementation of communication protocols and wireless terminals [7, 8]. Second, UML 2.0 and design tools provide formal action semantics and code generation, which enable rapid prototyping. Third, UML is an object-oriented language, and supports modular design approach that is an important aspect of reusable and flexible design.

This section presents briefly the main related work considering UML modeling in embedded systems design, and the parallel, and distributed execution of applications.

2.1. UML modeling with embedded systems

The UML profiles for the domain of real-time and embedded systems design can be roughly divided into three categories: system and platform design, performance modeling, and behavioral design. Next, the main related proposals are presented.

The *embedded UML* [9] is a UML profile proposal suitable for embedded real-time system specification, design, and verification. It represents a synthesis of concepts in hardare/software codesign. It presents extensions that define functional encapsulation and composition, communication specification, and mapping for performance evaluation.

A *UML platform profile* is proposed in [10], which presents a graphical language for the specification. It includes domain-specific classifiers and relationships to model the structure and behavior of embedded systems. The profile

introduces new building blocks to represent platform resources and services, and presents proper UML diagrams and notations to model platforms in different abstraction levels.

The *UML profile for schedulability, performance, and time* (or UML-SPT) is standardized by OMG [11]. The profile defines notations for building models of real-time systems with relevant quality of service (QoS) parameters. The profile supports the interoperability of modeling and analysis tools. However, it does not specify a full methodology, and the proile is considered to be very complex to utilize.

The *UML-RT profile* [12] defines execution semantics to capture behavior for simulation and synthesis. The profile presents capsules to represent system components, the internal behavior of which is designed with state machines. The capabilities to model architecture and performance are very limited in UML-RT, and thus, it should be considered complementary to the real-time UML profile. *HASoC* [13] is a design methodology that is based on UML-RT. It proposes also additional models of computation for the design of internal behavior.

In [14], Pllana and Fahringer present a set of building blocks to model concepts of message passing and shared memory. The proposed building blocks are parameterized to exploit time constructs in modeling. Further, they present an approach to map activity diagrams to process topologies.

OMG has recently introduced specifications for SoC and systems design domains. The *UML profile for SoC* [15] presents syntax for modeling modules and channels, the fundamental elements of SoC design. Further, the profile enables describing the behavior of a SoC using protocols and synchronicity semantics. The *OMG systems modeling language* (SysML) [16], and related *UML profile for systems engineering*, presents a new general-purpose modeling language for systems engineering. SysML uses a subset of UML, and its objective is to improve analysis capabilities.

These proposed UML profiles contain several features for utilizing UML in embedded and real-time domains. However, they are particularly targeted to single distinct aspects of design, and they miss the completeness for combining application and platform in an *implementation-oriented* fashion. It seems that many research activities have spent years and years for specifying astonishingly complex profiles that have only minor (reported) practical use.

2.2. Parallelism and distributed execution

Studies in microprocessor design have shown that a multiprocessor architecture consisting of several simple CPUs can outperform a single CPU using the same area [17] if the application has a large degree of parallelism. For the communications subsystem, Kaiserswerth has analyzed parallelism in communication protocols [18], stating that they are suitable for distributed execution, since they can be parallelized efficiently and also allow for pipelined execution.

Several parallel solutions have been developed to reduce the high computational complexity of video encoding [19]. *Temporal parallelism* [20, 21] exploits the independency between subsequent video frames. Consequently, the frame prediction is problematic because it limits the available parallelism. Furthermore, the induced latency may be intolerable in real-time systems. For *functional parallelism* [22–24], different functions are pipelined and executed in parallel on different processing units. This method is very straightforward and can efficiently exploit application-specific hardware accelerators. However, it may have limited scalability. In data parallelism [25, 26] video frames are divided into uniform spatial regions that are encoded in parallel. A typical approach is to use horizontal slice structures for this.

A common approach for simplifying the design of distributed systems is to utilize *middleware*, such as the *common object request broker architecture (CORBA)* [27], to abstract the underlying hardware for the application. OMG has also specified a *UML profile for CORBA*, which allows the presentation of CORBA semantics in UML [28]. However, the general middleware implementations are too complex for embedded systems. Thus, several lighter middleware approaches have been developed especially for real-time embedded systems [29–31]. However, Rintaluoma et al. [32] state that the overhead caused by the software layering and middleware have significant influence on performance in embedded multimedia applications.

In [33], Born et al. have presented a method for the design and development of distributed applications using UML. It uses automatic code generation to create code skeletons for component implementations on a middleware platform. Still, direct executable code generation from UML models, or modeling of hardware in UML, is not utilized.

2.3. Our approach

In this work, we use TUT-profile [34] that is a UML profile especially targeted to improve design efficiency and flexibility in the implementation and rapid prototyping of embedded real-time systems. The profile introduces a set of UML stereotypes which categorize and parameterize model constructs to enable extensive design automation both in analysis and implementation.

This work uses TUT-profile and the related design methodology in the design of parallel applications. The developed platforms and run-time environment seamlessly support functional parallelism and distributed execution of applications modeled in UML. The cost we have to pay for this is the overhead in execution time and increased memory usage. We argue that these drawbacks are highly accepted as we gain significant improvement in design efficiency.

The improved design efficiency comes from the clear modeling constructs and reduced amount of "low-level" coding, high-degree of design automation, easy model modifications and rapid prototyping, and improved design management and reuse. Unfortunately, these benefits in design efficiency are extremely hard to quantify, in contrast to the measurable overheads, but we will discuss our experiences in the design process.

None of the listed works provide fully automated design tools and practical, complex case studies on the deployment of the methods. To our best knowledge, the case study presented in this paper is the most complex design case that utilizes UML-based design automation for automated parallelization and distribution in this scale.

3. UML MODELING WITH KOSKI

In Koski, the whole design flow is governed by UML models designed according to a well-defined UML profile for embedded system design, called TUT-profile [34, 35]. The profile introduces a set of UML stereotypes which categorize and parameterize model elements to improve design automation both in analysis and implementation. The TUT-profile divides UML modeling into the design of *application*, *architecture*, and *mapping models*.

The application model is independent of hardware architecture and defines both the functionality and structure of an application. In a complex terminal with several subsystems, each subsystem can be described in a separate application model. In the TUT-profile, *application process* is an elementary unit of execution, which is implemented as an asynchronously communicating extended finite state machine (EFSM) using UML statecharts with action semantics [36, 37]. Further, existing library functions, for example DSP functions written in C, can be called inside the statecharts to enable efficient reuse.

The architecture model is independent of the application, and instantiates the required set of hardware components according to the needs of the current design. Hardware components are selected from a platform library that contains available processing elements as well as on-chip communication networks and interfaces for external (off-chip) devices. Processing elements are either general-purpose processors or dedicated hardware accelerators. The UML models of the components are abstract parameterized models, and do not describe the functionality.

The mapping model defines the mapping of an application to an architecture, that is, how application processes are executed on the instantiated processing elements. The mapping is performed in two stages. First, application processes are grouped into process groups. Second, the process groups are mapped to an architecture. Grouping can be performed according to different criteria, such as workload distribution and communication activity between groups. It should be noted that the mapping model is not compulsory. Koski tools perform the mapping automatically, but the designer can also control the mapping manually using the mapping model.

TUT-profile is further discussed below, in the implementation of the wireless video terminal.

3.1. Design flow and tools

Koski enables a fully automated implementation for a multiprocessor SoC on FPGA according to the UML models. A simplified view is presented in Figure 1. Koski comprises commercial design tools and self-made tools [38, 39] as presented in Table 1. A detailed description of the flow is given in [5].

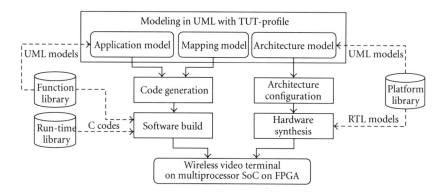

FIGURE 1: UML-based design flow for the implementation of the wireless video terminal.

TABLE 1: Categorization of the components and tools used in Koski.

Category	Self-made components/tools	Off-the-shelf components/tools
Application	TUTMAC UML model Video encoder UML model	
Design methodology and tools	TUT-profile Application distribution tool Architecture configuration tool Koski GUI Execution monitor	Tau G2 UML 2.0 tool Quartus II 5.1 Nios II GCC toolset
Software platform	IPC support functions HIBI API Hardware accelerator device drivers	eCos RTOS State machine scheduler
Hardware platform	HIBI communication architecture Nios-HIBI DMA Hardware accelerators Extension card for WLAN radio Extension card for on-board camera	Nios II softcore CPU FPGA development board Intersil WLAN radio transceiver OmniVision on-board camera module

Based on the application and mapping models, Koski generates code from UML statecharts, includes library functions and a run-time library, and finally builds distributed software implementing desired applications and subsystems on a given architecture. Based on the architecture model, Koski configures the library-based platform using the architecture configuration tool [38], and synthesizes the hardware for a multiprocessor SoC on FPGA.

4. EXECUTION PLATFORM

This section presents the execution platform including both the multiprocessor SoC platform and the software platform for the application distribution.

4.1. Hardware platform

The wireless video terminal is implemented on an Altera FPGA development board. The development board comprises Altera Stratix II EP2S60 FPGA, external memories (1 MB SRAM, 32 MB SDR SDRAM, 16 MB flash), and external interfaces (Ethernet and RS-232). Further, we have added

extension cards for a WLAN radio and on-board camera on the development board. The WLAN radio is Intersil MAC-less 2.4 GHz WLAN radio transceiver, which is compatible with the 802.11b physical layer, but does not implement the medium access control (MAC) layer. The on-board camera is OmniVision OV7670FSL camera and lens module, which features a single-chip VGA camera and image processor. The camera has a maximum frame rate of 30 fps in VGA and supports image sizes from VGA resolution down to 40×30 pixels. A photo of the board with the radio and camera cards is presented in Figure 2. The development board is connected to PC via Ethernet (for transferring data) and serial cable (for debug, diagnostics, and configuration).

The multiprocessor SoC platform is implemented on FPGA. The platform contains up to five Nios II processors; four processors for application execution, and one for debugging purposes and interfacing Ethernet with TCP/IP stack. With a larger FPGA device, such as Stratix II EP2S180, up to 15 processors can be used. Further, the platform contains dedicated hardware modules, such as hardware accelerators and interfaces to external devices [38]. These coarse-grain intellectual property (IP) blocks are connected using

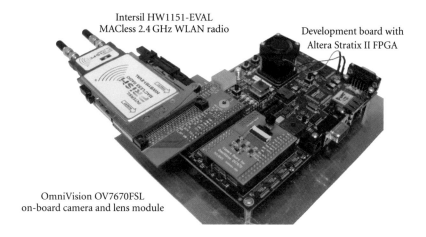

FIGURE 2: FPGA development board with the extension cards for WLAN radio and on-board camera.

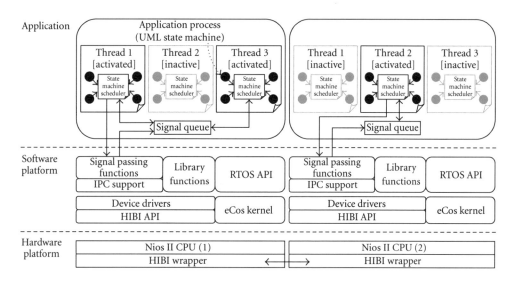

FIGURE 3: Structure of the software platform on hardware.

the heterogeneous IP block interconnection (HIBI) on-chip communication architecture [40]. Each processor module is self-contained, and contains a Nios II processor core, direct memory access (DMA) controller, timer units, instruction cache, and local data memory.

4.2. Software platform

The software platform enables the distributed execution of applications. It comprises the library functions and the run-time environment. The software platform on hardware is presented in Figure 3.

The library functions include various DSP and data processing functions (DCT, error checking, encryption) that can be used in the UML application models. In addition to the software-implemented algorithms, the library com-

prises software drivers to access their hardware accelerators and other hardware components, for example the radio interface.

The run-time environment consists of a real-time operating system (RTOS) application programming interface (API), interprocessor communication (IPC) support, state machine scheduler, and queues for signal passing between application processes. RTOS API implements thread creation and synchronization services through a standard interface. Consequently, different operating systems can be used on different CPUs. Currently, all CPUs run a local copy of eCos RTOS [41].

Distributed execution requires that information about the process mapping is included in the generated software. An application distributor tool parses this information automatically from the UML mapping model and creates the corresponding software codes. The codes include a mapping table that defines on which processing element each process group is to be executed.

4.2.1. Scheduling of application processes

When an RTOS is used, processes in the same process group of TUT-profile are executed in the same thread. The priority of the groups (threads) can be specified in the mapping model, and processes with real-time requirements can be placed in higher priority threads. The execution of processes within a thread is scheduled by an internal state machine scheduler. This schedule is nonpreemptive, meaning that state transitions cannot be interrupted by other transitions. The state machine scheduler is a library component, automatically generated by the UML tools.

Currently, the same generated program code is used for all CPUs in the system, which enables each CPU to execute all processes of the application. When a CPU starts execution, it checks the mapping table to decide which process groups (threads) it should activate; the rest of groups remains inactive on the particular CPU, as shown in Figure 3.

4.2.2. Signal passing for application processes

The internal (within a process group) and external (between process groups) signal passings are handled by signal passing functions. They take care that the signal is transmitted to the correct target process—regardless of the CPU the receiver is executed on and transparently to the application. The signal passing functions need services to transfer the UML signals between different processes. The IPC support provides services by negotiating the data transfers over the communication architecture and handling possible data fragmentation. On the lower layer, it uses the services of HIBI API to carry out the data transfers.

The signal passing at run-time is performed using two signal queues: one for signals passed inside the same thread and the other for signals from other threads. Processes within a thread share a common signal queue (included in state machine scheduler in Figure 3). When a signal is received, it is placed to the corresponding queue. When the state machine scheduler detects that a signal is sent to a process residing on a different CPU, the signal passing functions transmit the signal to the signal queue on the receiving CPU.

4.2.3. Dynamic mapping

The context of a UML process (state machine) is completely defined by its current state and the internal variables. Since all CPUs use the same generated program code, it is possible to remap processes between processing elements at run time without copying the application codes. Hence, the operation involves transferring only the process contexts and signals between CPUs, and updating the mapping tables.

Fast dynamic remapping is beneficial, for example, in power management, and in capacity management for applications executed in parallel on the same resources. During low load conditions, all processes can be migrated to single CPU and shut-down the rest. The processing power can be easily increased again when application load needs. Another benefit is the possibility to explore different mappings with real-time execution. This offers either speedup or accuracy gains compared to simulation-based or analytical exploration. The needed monitoring and diagnostic functionality are automatically included with Koski tools.

An initial version for automated remapping at run time according to workload is being evaluated. The current implementation observes the processor and workload statistics, and remaps the application processes to the minimum set of active processors. The implementation and results are discussed in detail in [42].

The dynamic mapping can be exploited also manually at run time using the execution monitor presented in Figure 4. The monitor shows the processors implemented on FPGA, application processes executed on the processors, and the utilization of each processor. A user can "drag-and-drop" processes from one processor to another to exploit dynamic mapping. In addition to the processor utilization, the monitor can show also other statistics, such as memory usage and bus utilization. Furthermore, application-specific diagnostic data can be shown, for example user data throughput in WLAN.

5. WIRELESS VIDEO TERMINAL

The wireless video terminal integrates two complementary subsystems: *video encoder* and *WLAN communications subsystems*. An overview of the wireless terminal is presented in Figure 5. In this section we present the subsystems and their UML application models, the hardware architecture and its UML architecture model, and finally, the mapping of subsystems to the architecture, and the corresponding UML mapping model.

The basic functionality of the terminal is as follows. The terminal receives raw image frames from PC over an Ethernet connection in IP packets, or from a camera directly connected to the terminal. The TCP/IP stack unwraps the raw frame data from the IP packets. The raw frame data is forwarded to the video encoder subsystem that produces the encoded bit stream. The encoded bit stream is forwarded to the communication subsystem that wraps the bit stream in WLAN packets and sends them over wireless link to a receiver.

The composite structure of the whole terminal is presented in Figure 6. This comprises the two subsystems and instantiates processes for bit stream packaging, managing TUTMAC, and accessing the external radio. The *bit stream packaging* wraps the encoded bit stream into user packets of TUTMAC. Class *MngUser* acts as a management instance that configures the TUTMAC protocol, that is, it defines the terminal type (base station or portable terminal), local station ID, and MAC address. *Radio* accesses the radio by configuring it and initiating data transmissions and receptions.

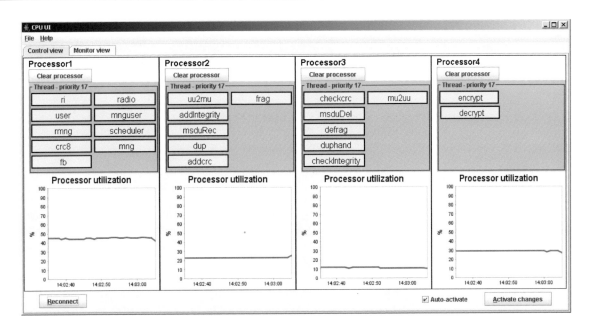

FIGURE 4: User interface of the execution monitor enables "drag-and-drop style" dynamic mapping.

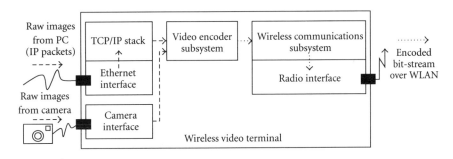

FIGURE 5: Overview of the wireless video terminal.

5.1. Video encoder subsystem

The video encoder subsystem implements an H.263 encoder in a function-parallel manner. Each function is implemented as a single UML process with well-defined interfaces.

As TUT-profile natively supports function parallelism, each process can be freely mapped to any (general-purpose) processing element even at run time. Further, the processes communicate using signals via their interfaces, and they have no shared (global) data.

The composite structure of the H.263 encoder UML model is presented in Figure 7. The application model for the encoder contains four processes. *Preprocessing* takes in frames of raw images and divides them into macroblocks. *Discrete cosine transformation* (DCT) transforms a macroblock into a set of spatial frequency coefficients. *Quantization* quantizes the coefficients. *Macroblock coding* (MBCoding) does entropy coding for macroblocks, and produces an encoded bit stream.

The functionality of the processes is obtained by reusing the C codes from a reference H.263 intraframe encoder. The control structure of the encoder was reimplemented using

UML statecharts, but the algorithms (DCT, quantization, coding) were reused as such. Thus, we were able to reuse over 90% of the reference C codes. The C codes for the algorithm implementations were added to the function library.

First stage in the modeling of the encoder was defining appropriate interfaces for the processes. For this, we defined data types in UML for frames, macroblocks, and bit stream, as presented in Figure 8(a). We chose to use C type of arrays (CArray) and pointers (CPtr) to store and access data, because in this way full compatibility with the existing algorithm implementations was achieved.

The control structures for the encoder were implemented using UML statecharts. Figure 8(b) presents the statechart implementation for the preprocessing. As mentioned before, the main task of the preprocessing is to divide frames into macroblocks. Further, the presented statechart implements flow control for the processing of created macroblocks. The flow control takes care that sufficient amount of macroblocks (five macroblocks in this case) is pipelined to the other encoder processes. This enables function-parallel processing as there are enough macroblocks in the pipeline. Also, this controls the size of signal queues as there are not too many

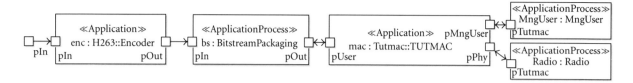

FIGURE 6: Top-level composite structure of the wireless video terminal.

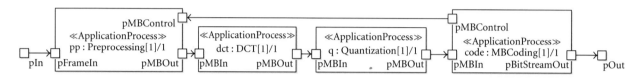

FIGURE 7: Composite structure of the video encoder.

macroblocks buffered within the processes, which increases dynamic memory usage.

5.2. WLAN communications subsystem

The WLAN communications subsystem implements a proprietary WLAN MAC protocol, called TUTMAC. It utilizes dynamic reservation time division multiple access (TDMA) to share the wireless medium [43]. TUTMAC solved the problems of scalability, QoS, and security present in standard WLANs. The wireless network has a centrally controlled topology, where one base station controls and manages multiple portable terminals. Several configurations have been developed for different purposes and platforms. Here we consider one configuration of the TUTMAC protocol.

The protocol contains data processing functions for cyclic redundancy check (CRC), encryption, and fragmentation. CRC is performed for headers with CRC-8 algorithm, and for payload data with CRC-32 algorithm. The encryption is performed for payload data using an advanced encryption system (AES) algorithm. The AES algorithm encrypts payload data in 128-bit blocks, and uses an encryption key of the same size. The fragmentation divides large user packets into several MAC frames. Further, processed frames are stored in a frame buffer. The TDMA scheduler picks the stored frames and transmits them in reserved time slots. The data processing is performed for every packet sent and received by a terminal. When the data throughput increases and packet interval decreases, several packets are pipelined and simultaneously processed by different protocol functions.

The TDMA scheduling has to maintain accurate frame synchronization. Tight real-time constraints are addressed and prioritized processing is needed to guarantee enough performance (throughput, latency) and accuracy (TDMA scheduling) for the protocol processing. Thus, the performance and parallel processing of protocol functions become significant issues. Depending on the implementation, the algorithms may also need hardware acceleration to meet the

delay bounds for data [39]. However, in this case we consider a full software implementation, because we want to emphasize the distributed software execution.

The top-level class composition of the TUTMAC protocol is presented in Figure 9(a). The top-level class (TUTMAC) introduces two processes and four classes with further composite structure, each introducing a number of processes, as presented in the hierarchical composite structure in Figure 9(b). Altogether, the application model of TUTMAC introduces 24 processes (state machines). The protocol functionality is fully defined in UML, and the target executables are obtained with automatic code generation. The implementation of the TUTMAC protocol using UML is described in detail in [7, 8].

5.3. Hardware architecture

The available components of the used platform are presented in a class diagram in Figure 10(a). The available components include different versions of Nios II (fast, standard economy [44], I/O with Ethernet), hardware accelerators (CRC32, AES), WLAN radio interface, and HIBI for on-chip communications. Each component is modeled as a class with an appropriate stereotype containing tagged values that parameterize the components (type, frequency). All processing elements have local memories and, hence, no memories are shown in the figure.

The architecture model for the wireless video terminal is presented in Figure 10(b). The architecture instantiates a set of components introduced by the platform. Further, it defines the communication architecture which, in this case, comprises one HIBI segment interconnecting the instantiated components.

5.4. Mapping of subsystems

As presented above, the subsystems of the terminal are modeled as two distinct applications. Further, these are integrated

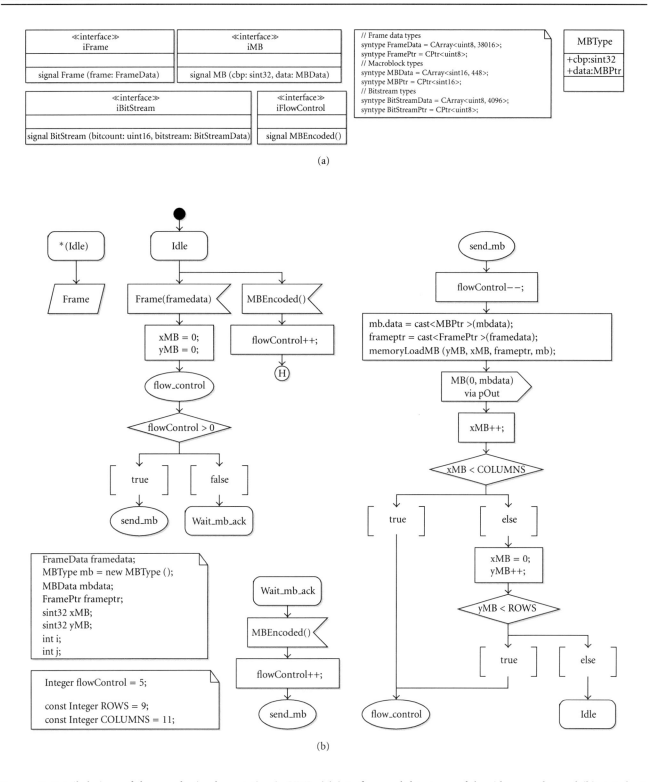

FIGURE 8: Detailed views of the encoder implementation in UML: (a) interfaces and data types of the video encoder, and (b) statechart implementation for the preprocessing.

together in a top-level application model that gathers the all functional components of the terminal.

Altogether, the terminal comprises 29 processes that are mapped to an architecture. One possible mapping model is presented in Figures 11(a) and 11(b). Each process is grouped to one of the eight process groups, each of which mapped to a processing element. Note that the presented mapping illustrates also the mapping of processes to

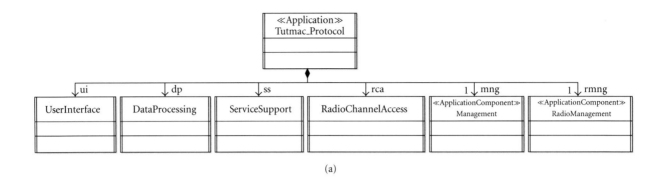

(a)

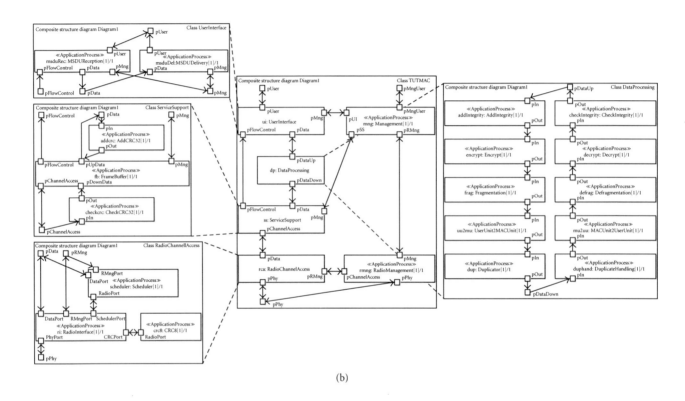

(b)

FIGURE 9: Hierarchical implementation of the TUTMAC protocol: (a) top-level class composition, and (b) hierarchical composite structure.

TABLE 2: Static memory requirements for a single CPU.

Software component	Code (bytes)	Code (%)	Data (bytes)	Data (%)	Total (bytes)	Total (%)
Generated code	28 810	20.52	56 376	43.59	85 186	31.58
Library functions	31 514	22.45	49 668	38.40	81 182	30.10
State machine scheduler	16 128	11.49	3 252	2.51	19 380	7.18
Signal passing functions	4 020	2.86	4	0.00	4 024	1.49
HIBI API	2 824	2.01	4 208	3.25	7 032	2.61
IPC support	2 204	1.57	449	0.35	2 653	0.98
Device drivers	1 348	0.96	84	0.06	1 432	0.53
eCos	53 556	38.14	15 299	11.83	68 855	25.53
Total software	140 404	100.00	129 340	100.00	269 744	100.00

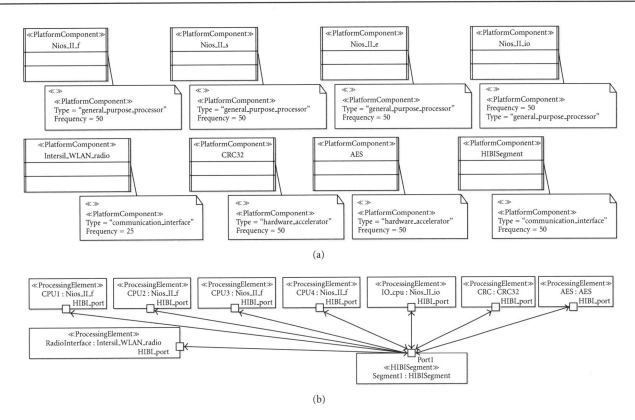

FIGURE 10: Platform components are (a) modeled as UML classes and parameterized using appropriate stereotypes, and (b) instantiated in the architecture model.

TABLE 3: Average processing times of the TUTMAC components for a single frame.

Component	Processing time (ms)	Processing time (%)	Note
msduRec	3.63	13.13	—
addIntegrity	0.94	3.38	—
encrypt	14.56	52.61	—
frag	0.66	2.39	—
uu2mu	4.10	14.80	(1)
addcrc	2.17	7.85	(1)
fb	0.64	2.30	(1)
ri	0.78	2.81	(1)
crc8	0.20	0.72	(1)
Total	27.68	100.00	—

(1) Processing time is for 2 WLAN packets (data is fragmented).

TABLE 4: Average processing times of the video encoder components for a single frame.

Component	Processing time (ms)	Processing time (%)
Preprocessing	17.83	9.31
DCT	46.93	24.51
Quantization	68.05	35.55
MBCoding	57.86	30.23
BitstreamPackaging	0.77	0.40
Total	191.43	100.00

6.1. Implementation details

The required amount of memory for each software component is presented in Table 2. All CPUs functionally have identical software images that differ in memory and process mappings only. Creating unique code images for each CPU was not considered at this stage of research. However, it is a viable option, especially, when the dynamic run-time remapping is not needed. In addition to the static memory needs, the applications require 140–150 kB of dynamic memory. The dynamic memory consumption is distributed among CPUs according to processes mapping.

The size of the hardware architecture (five Nios II CPUs, HIBI, radio interface, AES, CRC-32) is 20 495 adaptive logic

hardware accelerator, although in this case study we use full software implementation to concentrate the distributed execution of software.

6. MEASUREMENTS

This section presents the implementation details and performance measurements of the wireless video terminal.

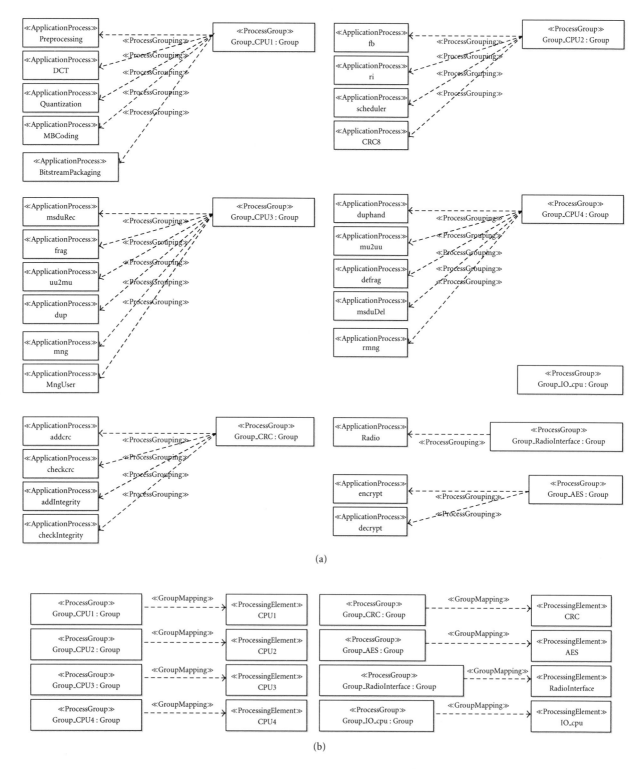

FIGURE 11: Mapping models: (a) grouping of processes to the process groups, and (b) mapping of process groups to the architecture.

modules (ALM), which takes 84% of the total capacity on the used Stratix II FPGA. Further, the hardware (FIFO modules, configuration bits) takes 760 kb (29%) of the FPGA on-chip memory. The operating frequency was set to 50 MHz in all measurements.

6.2. Performance measurements

Table 3 presents the average processing times of the TUT-MAC components when transmitting a single encoded video frame. The size of the encoded bit stream per frame was

TABLE 5: Video encoder frame rates and TUTMAC transmission delay with different mappings.

Mapping	Frame rate (fps)	Transmission delay (ms)	Note
Whole terminal on a single CPU	1.70	54.10	(1)
Video encoder on 1 CPU, TUTMAC on 1 CPU	2.10	27.70	—
Video encoder on 2 CPUs, TUTMAC on 1 CPU	2.40	27.70	—
Video encoder on 3 CPUs, TUTMAC on 1 CPU	3.00	27.70	—
Video encoder on 4 CPUs, TUTMAC on 1 CPU	2.80	28.00	(1)

(1) One CPU is shared.

1800 B in average. The maximum WLAN packet size is 1550 B, which means that each encoded frame was fragmented into two WLAN packets. The total processing time (27.68 ms) in this case results in a theoretical maximum throughput of 500 kbps, which is very adequate to transmit the encoded bit stream.

The processing times for the encoder components are given in Table 4. DCT, quantization, and macroblock coding handle frames in macroblocks. The presented values are total times for a single frame (11×9 macroblocks). The total processing time (191.43 ms) results in a theoretical maximum frame rate of 5.2 fps on a single CPU (no parallelization). The reference C implementation of the encoder (on which the UML implementation is based) achieved the frame rate of 4.5 fps on the same hardware.

The presented times include only computing, but not communication between processes. The run time overheads of interprocess communication are currently being evaluated and optimized. AES and CRC that constitute over 60% of the frame processing time could also be executed on hardware accelerator. DCT and motion estimation accelerators for video encoding are currently being integrated.

The frame rate of the video encoder and the transmission delay of TUTMAC were measured with different mappings. According to the results presented above, we decided to concentrate on the distribution of the video encoder, because it requires more computing effort. Further, TUTMAC is assumed to operate well on a single CPU as the data throughput is rather low (only few dozen kbps).

The frame rates and transmission delays with different mappings are shown in Table 5. In the first case, the whole terminal was mapped to a single CPU. In the second case, the video encoder and the TUTMAC protocol were executed on separate CPUs. In the third and fourth cases, the video encoder was distributed into two and three CPUs, respectively, while the TUTMAC protocol was executed on one CPU. Finally, in the fifth case, the video encoder was executed on both four CPUs and TUTMAC shared one of the four CPUs. As mentioned before, remapping does not require hardware synthesis, or even software compilation.

The measurements revealed that the distributed execution of the video encoder improves the frame rate, and at the most, the frame rate is 3.0 fps on three CPUs. In the fifth case, the sharing of one CPU increases the workload on that CPU, which prevents further improvements in frame rate.

The communication overhead between CPUs is the main reason of the fact that the improvements are lower than in

an ideal case. However, we argue that the achieved results in performance are very good as the used design methodology and tools improve the design efficiency significantly. It should also be noted that the video encoder is not processor optimized but is based on fully portable models.

7. CONCLUSIONS

This paper presented the implementation of a wireless video terminal using UML-based design flow. The terminal comprises a function parallel H.263 video encoder and WLAN subsystem for wireless communications. The whole terminal, including the application and platform, was modeled in UML, and full design automation was used to the physical implementation.

The main objective of this work was to study the feasibility of the used design methodology and tools to implement a multimedia terminal comprising various subsystems, each comprises several functional components. This objective was fulfilled with very pleasant results as the design flow tools enable extensive design automation in implementation from high-abstraction level models to a complete multiprocessor SoC on FPGA. The experiments with the UML-based design flow proved its suitability and competence in designing also complex embedded multimedia terminals.

The performance of the video encoding was quite satisfactory as we achieved 3.0 fps without any optimizations in architecture and communications. Slightly better performance can be achieved using reference C implementation of the encoder. The reduced performance is the cost of using UML and design automation, but is highly accepted as we gain significant improvement in design efficiency.

Capability to rapid prototyping and easy modifications to the applications is one of the major improvements in the design process as the fully automated design flow significantly reduces the amount of "low-level" coding. Further, the clear constructs in modeling, due to the well-defined and practical profile, enable rather easy integration of complex subsystems, as shown in this case study.

The future work with the design methodology includes enhanced support for nonfunctional constraints and more detailed hardware modeling. In addition, IPC functions and memory architecture will be optimized to allow more efficient parallelization. The encoder could be implemented in data or temporal parallel fashion to enhance the scalability and performance. Further, the application development will include the implementation of the full H.263/MPEG-4 encoder, that is, adding the motion estimation functionality to enable encoding the interframes also.

REFERENCES

[1] L. Lavagno, G. Martin, and B. Selic, Eds., *UML for Real: Design of Embedded Real-Time Systems*, Kluwer Academic, New York, NY, USA, 2003.

[2] R. Soley, "Model Driven Architecture," November 2000, Object Management Group (OMG), white paper.

[3] R. B. France, S. Ghosh, T. Dinh-Trong, and A. Solberg, "Model-driven development using UML 2.0: promises and pitfalls," *Computer*, vol. 39, no. 2, pp. 59–66, 2006.

[4] Object Management Group (OMG), "MDA Guide (Version 1.0.1)," June 2003.

[5] T. Kangas, P. Kukkala, H. Orsila, et al., "UML-based multiprocessor SoC design framework," *ACM Transactions on Embedded Computing Systems*, vol. 5, no. 2, pp. 281–320, 2006.

[6] Object Management Group (OMG), "Unified Modeling Language (UML) Superstructure Specification (Version 2.0)," August 2005.

[7] P. Kukkala, V. Helminen, M. Hännikäinen, and T. D. Hämäläinen, "UML 2.0 implementation of an embedded WLAN protocol," in *Proceedings of the 15th IEEE International Symposium on Personal, Indoor and Mobile Radio Communications (PIMRC '04)*, vol. 2, pp. 1158–1162, Barcelona, Spain, September 2004.

[8] P. Kukkala, M. Hännikäinen, and T. D. Hämäläinen, "Design and implementation of a WLAN terminal using UML 2.0 based design flow," in *Embedded Computer Systems: Architectures, Modeling, and Simulation*, vol. 3553 of *Lecture Notes in Computer Science*, pp. 404–413, Springer, New York, NY, USA, 2005.

[9] G. Martin, L. Lavagno, and J. Louis-Guerin, "Embedded UML: a merger of real-time UML and co-design," in *Proceedings of the 9th International Workshop Hardware/Software Codesign*, pp. 23–28, Copenhagen, Denmark, April 2001.

[10] R. Chen, M. Sgroi, L. Lavagno, G. Martin, A. Sangiovanni-Vincentelli, and J. Rabaey, "UML and platform-based design," in *UML for Real: Design of Embedded Real-Time Systems*, pp. 107–126, Kluwer Academic, Norwell, Mass, USA, 2003.

[11] Object Management Group (OMG), "UML Profile for Schedulability, Performance, and Time Specification (Version 1.1)," January 2005.

[12] B. Selic, "Using UML for modeling complex real-time systems," in *Proceedings of Languages, Compilers, and Tools for Embedded Systems (LCTES '98)*, vol. 1474 of *Lecture Notes in Computer Science*, pp. 250–260, Montreal, Canada, June 1998.

[13] P. Green, M. Edwards, and S. Essa, "HASoC - towards a new method for system-on-a-chip development," *Design Automation for Embedded Systems*, vol. 6, no. 4, pp. 333–353, 2002.

[14] S. Pllana and T. Fahringer, "On customizing the UML for modeling performance-oriented applications," in *Proceedings of the 5th International Conference on the Unified Modeling Language*, vol. 2460 of *Lecture Notes in Computer Science*, pp. 259–274, Springer, Dresden, Germany, September-October 2002.

[15] Object Management Group (OMG), "UML Profile for System on a Chip (SoC) Specification (Version 1.0)," June 2006.

[16] Object Management Group (OMG), "OMG Systems Modeling Language (SysML) Specification," June 2006.

[17] K. Olukotun, B. A. Nayfeh, L. Hammond, K. Wilson, and K. Chang, "The case for a single-chip multiprocessor," in *Proceed-

[18] M. Kaiserswerth, "The parallel protocol engine," *IEEE/ACM Transactions on Networking*, vol. 1, no. 6, pp. 650–663, 1993.

[19] I. Ahmad, Y. He, and M. L. Liou, "Video compression with parallel processing," *Parallel Computing*, vol. 28, no. 7-8, pp. 1039–1078, 2002.

[20] I. Agi and R. Jagannathan, "A portable fault-tolerant parallel software MPEG-1 encoder," *Multimedia Tools and Applications*, vol. 2, no. 3, pp. 183–197, 1996.

[21] J. Nang and J. Kim, "Effective parallelizing scheme of MPEG-1 video encoding on ethernet-connected workstations," in *Proceedings of the Conference on Advances in Parallel and Distributed Computing*, pp. 4–11, Shanghai, China, March 1997.

[22] M. J. Garrido, C. Sanz, M. Jiménez, and J. M. Menasses, "An FPGA implementation of a flexible architecture for H.263 video coding," *IEEE Transactions on Consumer Electronics*, vol. 48, no. 4, pp. 1056–1066, 2002.

[23] O. Cantineau and J.-D. Legat, "Efficient parallelisation of an MPEG-2 codec on a TMS320C80 video processor," in *Proceedings of the International Conference on Image Processing (ICIP '98)*, vol. 3, pp. 977–980, Chicago, Ill, USA, October 1998.

[24] S. Bhattacharjee, S. Das, D. Saha, D. R. Chowdhury, and P. P. Chaudhuri, "A parallel architecture for video compression," in *Proceedings of the 10th IEEE International Conference on VLSI Design*, pp. 247–252, Hyderabad, India, January 1997.

[25] S. M. Akramullah, I. Ahmad, and M. L. Liou, "Performance of software-based MPEG-2 video encoder on parallel and distributed systems," *IEEE Transactions on Circuits and Systems for Video Technology*, vol. 7, no. 4, pp. 687–695, 1997.

[26] N. H. C. Yung and K.-K. Leung, "Spatial and temporal data parallelization of the H.261 video coding algorithm," *IEEE Transactions on Circuits and Systems for Video Technology*, vol. 11, no. 1, pp. 91–104, 2001.

[27] Object Management Group (OMG), "The Common Object Request Broker Specification (Version 3.0)," March 2004.

[28] Object Management Group (OMG), "UML Profile for CORBA Specification (Version 1.0)," April 2002.

[29] D. C. Schmidt and F. Kuhns, "An overview of the real-time CORBA specification," *Computer*, vol. 33, no. 6, pp. 56–63, 2000.

[30] U. Brinkschulte, T. Ungerer, A. Bechina, et al., "A microkernel middleware architecture for distributed embedded real-time systems," in *Proceedings of the 20th IEEE Symposium on Reliable Distributed Systems (SRDS '01)*, pp. 218–226, New Orleans, La, USA, October 2001.

[31] C. Gill, V. Subramonian, J. Parsons, et al., "ORB middleware evolution for networked embedded systems," in *Proceedings of the 8th International Workshop on Object-Oriented Real-Time Dependable Systems (WORDS '03)*, pp. 169–176, Guadalajara, Mexico, January 2003.

[32] T. Rintaluoma, O. Silven, and J. Raekallio, "Interface overheads in embedded multimedia software," in *Proceedings of the 6th International Workshop on Architectures, Modeling, and Simulation (SAMOS '06)*, vol. 4017 of *Lecture Notes in Computer Science*, pp. 5–14, Springer, Samos, Greece, July 2006.

[33] M. Born, E. Holz, and O. Kath, "A method for the design and development of distributed applications using UML," in *Proceedings of the 37th International Conference on Technology of Object-Oriented Languages and Systems (TOOLS-PACIFIC '00)*, pp. 253–264, Sydney, Australia, November 2000.

ings of the 7th International Symposium on Architectural Support for Programming Languages and Operating Systems (ASPLOS '96), pp. 2–11, Cambridge, Mass, USA, October 1996.

[34] P. Kukkala, J. Riihimäki, M. Hännikäinen, T. D. Hämäläinen, and K. Kronlöf, "UML 2.0 profile for embedded system design," in *Proceedings of Design, Automation and Test in Europe (DATE '05)*, vol. 2, pp. 710–715, Munich, Germany, March 2005.

[35] P. Kukkala, M. Hännikäinen, and T. D. Hämäläinen, "Performance modeling and reporting for the UML 2.0 design of embedded systems," in *Proceedings of International Symposium on System-on-Chip (SoC '05)*, pp. 50–53, Tampere, Finland, November 2005.

[36] M. Björkander, "Graphical programming using UML and SDL," *Computer*, vol. 33, no. 12, pp. 30–35, 2000.

[37] S. Gnesi, D. Latella, and M. Massink, "Modular semantics for a UML statechart diagrams kernel and its extension to multi-charts and branching time model-checking," *Journal of Logic and Algebraic Programming*, vol. 51, no. 1, pp. 43–75, 2002.

[38] T. Arpinen, P. Kukkala, E. Salminen, M. Hännikäinen, and T. D. Hämäläinen, "Configurable multiprocessor platform with RTOS for distributed execution of UML 2.0 designed applications," in *Proceedings of Design, Automation and Test in Europe (DATE '06)*, vol. 1, pp. 1–6, Munich, Germany, March 2006.

[39] M. Setälä, P. Kukkala, T. Arpinen, M. Hännikäinen, and T. D. Hämäläinen, "Automated distribution of UML 2.0 designed applications to a configurable multiprocessor platform," in *Proceedings of the 6th International Workshop on Architectures, Modeling, and Simulation (SAMOS '06)*, vol. 4017 of *Lecture Notes in Computer Science*, pp. 27–38, Springer, 2006.

[40] E. Salminen, T. Kangas, T. D. Hämäläinen, J. Riihimäki, V. Lahtinen, and K. Kuusilinna, "HIBI communication network for system-on-chip," *Journal of VLSI Signal Processing Systems for Signal, Image, and Video Technology*, vol. 43, no. 2-3, pp. 185–205, 2006.

[41] A. Massa, *Embedded Software Development with eCos*, Prentice-Hall Professional Technical Reference, New York, NY, USA, 2002.

[42] P. Kukkala, T. Arpinen, M. Setälä, M. Hännikäinen, and T. D. Hämäläinen, "Dynamic power management for UML modeled applications on multiprocessor SoC," in *Proceedings of the IS&T/SPIE 19th Annual Symposium on Electronic Imaging*, San Jose, Calif, USA, January-February 2007.

[43] M. Hännikäinen, T. Lavikko, P. Kukkala, and T. D. Hämäläinen, "TUTWLAN - QoS supporting wireless network," *Telecommunication Systems*, vol. 23, no. 3-4, pp. 297–333, 2003.

[44] Altera, "Nios II Processor Reference Handbook (Version 6.0)," May 2006.

Hindawi Publishing Corporation
EURASIP Journal on Embedded Systems
Volume 2007, Article ID 75947, 13 pages
doi:10.1155/2007/75947

Research Article

pn: A Tool for Improved Derivation of Process Networks

Sven Verdoolaege, Hristo Nikolov, and Todor Stefanov

Leiden Institute of Advanced Computer Science (LIACS), Leiden University, Niels Bohrweg 1, 2333 CA, Leiden, The Netherlands

Received 30 June 2006; Revised 12 December 2006; Accepted 10 January 2007

Recommended by Shuvra Bhattacharyya

Current emerging embedded System-on-Chip platforms are increasingly becoming multiprocessor architectures. System designers experience significant difficulties in programming these platforms. The applications are typically specified as sequential programs that do not reveal the available parallelism in an application, thereby hindering the efficient mapping of an application onto a parallel multiprocessor platform. In this paper, we present our compiler techniques for facilitating the migration from a sequential application specification to a *parallel* application specification using the process network model of computation. Our work is inspired by a previous research project called Compaan. With our techniques we address optimization issues such as the generation of process networks with simplified topology and communication without sacrificing the process networks' performance. Moreover, we describe a technique for compile-time memory requirement estimation which we consider as an important contribution of this paper. We demonstrate the usefulness of our techniques on several examples.

1. INTRODUCTION AND MOTIVATION

The complexity of embedded multimedia and signal processing applications has reached a point where the performance requirements of these applications can no longer be supported by embedded system platforms based on a single processor. Therefore, modern embedded System-on-Chip platforms have to be multiprocessor architectures. In recent years, a lot of attention has been paid to building such multiprocessor platforms. Fortunately, advances in chip technology facilitate this activity. However, less attention has been paid to compiler techniques for efficient programming of multiprocessor platforms, that is, the efficient mapping of applications onto these platforms is becoming a key issue. Today, system designers experience significant difficulties in programming multiprocessor platforms because the way an application is specified by an application developer does not match the way multiprocessor platforms operate. The applications are typically specified as sequential programs using imperative programming languages such as C/C++ or Matlab. Specifying an application as a sequential program is relatively easy and convenient for application developers, but the sequential nature of such a specification does not reveal the available parallelism in an application. This fact makes the efficient mapping of an application onto a parallel multiprocessor platform very difficult. By contrast, if an application is specified using a parallel model of computation (MoC), then the mapping can be done in a systematic and transparent way using a disciplined approach [1]. However, specifying an application using a parallel MoC is difficult, not well understood by application developers, and a time consuming and error prone process. That is why application developers still prefer to specify an application as a sequential program, which is well understood, even though such a specification is not suitable for mapping an application onto a parallel multiprocessor platform.

This gap between a sequential program and a parallel model of computation motivates us to investigate and develop compiler techniques that facilitate the migration from a sequential application specification to a parallel application specification. These compiler techniques depend on the parallel model of computation used for parallel application specification. Although many parallel models of computation exist [2, 3], in this paper we consider the process network model of computation [4] because its operational semantics are simple, yet general enough to conveniently specify *stream-oriented* data processing that fits nicely with the application domain we are interested in—multimedia and signal processing applications. Moreover, for this application domain, many researchers [5–14] have already indicated that process networks are very suitable for systematic and efficient mapping onto multiprocessor platforms.

In this paper, we present our compiler techniques for deriving process network specifications for applications specified as static affine nested loop programs (SANLPs), thereby bridging the gap mentioned above in a particular way. SANLPs are important in scientific, matrix computation and multimedia and adaptive signal processing applications. Our work is inspired by previous research on Compaan [15–17]. The techniques presented in this paper and implemented in the pn tool of our isa tool set can be seen as a significant improvement of the techniques developed in the Compaan project in the following sense. The Compaan project has identified the fundamental problems that have to be solved in order to derive process networks systematically and automatically and has proposed and implemented basic solutions to these problems. However, many optimization issues that improve the quality of the derived process networks have not been fully addressed in Compaan. Our techniques try to address optimization issues in four main aspects.

Given an application specified as an SANLP,

(1) *derive (if possible) process networks (PN) with fewer communication channels between different processes compared to Compaan-derived PNs without sacrificing the PN performance;*

(2) *derive (if possible) process networks (PN) with fewer processes compared to Compaan-derived PNs without sacrificing the PN performance;*

(3) *replace (if possible) reordering communication channels with simple FIFO channels without sacrificing the PN performance;*

(4) *determine the size of the communication FIFO channels at compile time.* The problem of deriving efficient FIFO sizes has not been addressed by Compaan. Our techniques for computing FIFO sizes constitute a starting point to overcome this problem.

2. RELATED WORK

The work in [11] presents a methodology and techniques implemented in a tool called ESPAM for automated multiprocessor system design, programming, and implementation. The ESPAM design flow starts with three input specifications at the system level of abstraction, namely a platform specification, a mapping specification, and an application specification. ESPAM requires the application specification to be a process network. Our compiler techniques presented in this paper are primarily intended to be used as a font-end tool for ESPAM. (Kahn) process networks are also supported by the Ptolemy II framework [3] and the YAPI environment [5] for concurrent modeling and design of applications and systems. In many cases, manually specifying an application as a process network is a very time consuming and error prone process. Using our techniques as a front-end to these tools can significantly speedup the modeling effort when process networks are used and avoid modeling errors because our techniques guarantee a correct-by-construction generation of process networks.

Process networks have been used to model applications and to explore the mapping of these applications onto multiprocessor architectures [6, 9, 12, 14]. The application modeling is performed manually starting from sequential C code and a significant amount of time (a few weeks) is spent by the designers on correctly transforming the sequential C code into process networks. This activity slows down the design space exploration process. The work presented in this paper gives a solution for fast automatic derivation of process networks from sequential C code that will contribute to faster design space exploration.

The relation of our analysis to Compaan will be highlighted throughout the text. As to memory size requirements, much research has been devoted to optimal reuse of memory for arrays. For an overview and a general technique, we refer to [18]. These techniques are complementary to our research on FIFO sizes and can be used on the reordering channels and optionally on the data communication inside a node. Also related is the concept of reuse distances [19]. In particular, our FIFO sizes are a special case of the "reuse distance per statement" of [20]. For more advanced forms of copy propagation, we refer to [21].

The rest of this paper is organized as follows. In Section 3, we first introduce some concepts that we will need throughout this paper. We explain how to derive and optimize process networks in Section 4 and how to compute FIFO sizes in Section 5. Detailed examples are given in Section 6, with a further comparison to Compaan-generated networks in Section 7. In Section 8, we conclude the paper.

3. PRELIMINARIES

In this section, we introduce the process network model, discuss static affine nested loop programs (SANLPs) and our internal representation, and introduce our main analysis tools.

3.1. *The process network model*

As the name suggests, a process network consists of a set of *processes*, also called *nodes*, that communicate with each other through *channels*. Each process has a fixed internal schedule, but there is no (a priori) global schedule that dictates the relative order of execution of the different processes. Rather, the relative execution order is solely determined by the channels through which the processes communicate. In particular, a process will block if it needs data from a channel that is not available yet. Similarly, a process will block if it tries to write to a "full" channel.

In the special case of a Kahn process network (KPN), the communication channels are unbounded FIFOs that can only block on a read. In the more general case, data can be written to a channel in an order that is different from the order in which the data is read. Such channels are called *reordering channels*. Furthermore, the FIFO channels have additional properties such as their size and the ability to be implemented as a shift register. Since both reads and writes may block, it is important to ensure the FIFOs are large enough to avoid deadlocks. Note that determining suitable channel sizes may not be possible in general, but it is possible for process networks derived from SANLPs as defined in Section 3.2. Our networks can be used as input for tools that expect Kahn process networks by ignoring the additional properties of FIFO channels and by changing the order in which a process reads from a reordering channel to match the order of the writes and storing the data that is not needed yet in an internal memory block.

3.2. Limitations on the input and internal representation

The SANLPs are programs or program fragments that can be represented in the well-known polytope model [22]. That is, an SANLP consists of a set of statements, each possibly enclosed in loops and/or guarded by conditions. The loops need not be perfectly nested. All lower and upper bounds of the loops as well as all expressions in conditions and array accesses can contain enclosing loop iterators and parameters as well as modulo and integer divisions, but no products of these elements. Such expressions are called quasi-affine. The parameters are symbolic constants, that is, their values may not change during the execution of the program fragment. These restrictions allow a compact representation of the program through sets and relations of integer vectors defined by linear (in)equalities, existential quantification, and the union operation. More technically, our (parametric) "integer sets" and "integer relations" are (disjoint) unions of projections of the integer points in (parametric) polytopes.

In particular, the set of iterator vectors for which a statement is executed is an integer set called the *iteration domain*. The linear inequalities of this set correspond to the lower and upper bounds of the loops enclosing the statement. For example, the iteration domain of statement S1 in Figure 1 is $\{i \mid 0 \le i \le N - 1\}$. The elements in these sets are ordered according to the order in which the iterations of the loop nest are executed, assuming the loops are normalized to have step +1. This order is called the *lexicographical order* and will be denoted by $\prec$. A vector $\mathbf{a} \in \mathbb{Z}^n$ is said to be lexicographically (strictly) smaller than $\mathbf{b} \in \mathbb{Z}^n$ if for the first position i in which $\mathbf{a}$ and $\mathbf{b}$ differ, we have $a_i < b_i$, or, equivalently,

$$\mathbf{a} \prec \mathbf{b} \equiv \bigvee_{i=1}^{n} \left(a_i < b_i \wedge \bigwedge_{j=1}^{i-1} a_j = b_j \right). \quad (1)$$

The iteration domains will form the basis of the description of the nodes in our process network, as each node will correspond to a particular statement. The channels are determined by the array (or scalar) accesses in the corresponding statements. All accesses that appear on the left-hand side of an assignment or in an address-of (&) expression are considered to be *write accesses*. All other accesses are considered to be *read accesses*. Each of these accesses is represented by an *access relation*, relating each iteration of the statement to the array element accessed by the iteration, that is, $\{(\mathbf{i}, \mathbf{a}) \in I \times A \mid \mathbf{a} = L\mathbf{i} + \mathbf{m}\}$, where I is the iteration domain, A is the array space, and $L\mathbf{i} + \mathbf{m}$ is the affine access function.

The use of access *relations* allows us to impose additional constraints on the iterations where the access occurs.

```
for (i = 0; i < N; ++i)
S1: b[i] = f(i > 0 ? a[i-1] : a[i], a[i],
              i < N-1 ? a[i+1] : a[i]);
for (i = 0; i < N; ++i) {
    if (i > 0)
        tmp = b[i-1];
    else
        tmp = b[i];
S2: c[i] = g(b[i], tmp);
}
```

FIGURE 1: Use of temporary variables to express border behavior.

This is useful for expressing the effect of the ternary operator (?:) in C, or, equivalently, the use of temporary scalar variables. These frequently occur in multimedia applications where one or more kernels uniformly manipulate a stream of data such as an image, but behave slightly differently at the borders. An example of both ways of expressing border behavior is shown in Figure 1 on a 1D data stream. The second read access through b in line 9, after elimination of the temporary variable tmp, can be represented as

$$R = \{(i, a) \mid a = i - 1 \wedge 1 \le i \le N - 1\} \cup \{(i, a) \mid a = i = 0\}. \quad (2)$$

To eliminate such temporary variables, we first identify the statements that simply copy data to a temporary variable, perform a dataflow analysis (as explained in Section 4.1) on those temporary variables in a first pass and combine the resulting constraints with the access relation from the copy statement. A straightforward transformation of code such as that of Figure 1 would introduce extra nodes that simply copy the data from the appropriate channel to the input channel of the core node. Not only does this result in a network with more nodes than needed, it also reduces the opportunity for reducing internode communication.

3.3. Analysis tools: lexicographical minimization and counting

Our main analysis tool is parametric integer programming [23], which computes the lexicographically smallest (or largest) element of a parametric integer set. The result is a subdivision of the parameter space with for each cell of this subdivision a description of the corresponding unique minimal element as an affine combination of the parameters and possibly some additional existentially quantified variables. This result can be described as a union of parametric integer sets, where each set in the union contains a single point, or alternatively as a relation, or indeed a function, between (some of) the parameters and the corresponding lexicographical minimum. The existentially quantified variables that may appear will always be uniquely quantified, that is, the existential quantifier $\exists$ is actually a uniqueness quantifier $\exists!$. Parametric integer programming (PIP) can be used to project out some of the variables in a set. We simply compute the lexicographical minimum of these variables, treating all other variables as additional parameters, and then discard the description of the minimal element.

The `barvinok` library [24] efficiently computes the number of integer points in a parametric polytope. We can use it to compute the number of points in a parametric set provided that the existentially quantified variables are uniquely quantified, which can be ensured by first using PIP if needed. The result of the computation is a compact representation of a function from the parameters to the nonnegative integers, the number of elements in the set for the corresponding parameter values. In particular, the result is a piecewise quasipolynomial in the parameters. The `bernstein` library [25] can be used to compute an upper bound on a piecewise polynomial over a parametric polytope.

4. DERIVATION OF PROCESS NETWORKS

This section explains the conversion of SANLPs to process networks. We first derive the channels using a modified dataflow analysis in Section 4.1 and then we show how to determine channel types in Section 4.2 and discuss some optimizations on self-loop channels in Section 4.3.

4.1. Dataflow analysis

To compute the channels between the nodes, we basically need to perform array dataflow analysis [26]. That is, for each execution of a read operation of a given data element in a statement, we need to find the source of the data, that is, the corresponding write operation that wrote the data element. However, to reduce communication between different nodes and in contrast to standard dataflow analysis, we will also consider all previous read operations from the same statement as possible sources of the data.

The problem to be solved is then: given a read from an array element, what was the last write to or read (from that statement) from that array element? The last iteration of a statement satisfying some constraints can be obtained using PIP, where we compute the lexicographical *maximum* of the write (or read) source operations in terms of the iterators of the "sink" read operation. Since there may be multiple statements that are potential sources of the data and since we also need to express that the source operation is executed before the read (which is not a linear constraint, but rather a disjunction of n linear constraints (1), where n is the shared nesting level), we actually need to perform a number of PIP invocations. For details, we refer to [26], keeping in mind that we consider a larger set of possible sources.

For example, the first read access in statement S2 of the code in Figure 1 reads data written by statement S1, which results in a channel from node "S1" to node "S2." In particular, data flows from iteration i_w of statement S1 to iteration $i_r = i_w$ of statement S2. This information is captured by the integer relation

$$D_{\text{S1}\rightarrow\text{S2}} = \{(i_w, i_r) \mid i_r = i_w \wedge 0 \le i_r \le N-1\}. \quad (3)$$

For the second read access in statement S2, as described by (2), the data has already been read by the same statement after it was written. This results in a self-loop from S2 to itself described as

$$\begin{aligned} D_{\text{S2}\rightarrow\text{S2}} = \{(i_w, i_r) \mid i_w = i_r - 1 \wedge 1 \le i_r \le N-1\} \\ \cup \{(i_w, i_r) \mid i_w = i_r = 0\}. \end{aligned} \quad (4)$$

In general, we obtain pairs of write/read and read operations such that some data flows from the write/read operation to the (other) read operation. These pairs correspond to the channels in our process network. For each of these pairs, we further obtain a union of integer relations

$$\bigcup_{j=1}^{m} D_j(\mathbf{i}_w, \mathbf{i}_r) \subset \mathbb{Z}^{n_1} \times \mathbb{Z}^{n_2}, \quad (5)$$

with n_1 and n_2 the number of loops enclosing the write and read operation respectively, that connect the specific iterations of the write/read and read operations such that the first is the source of the second. As such, each iteration of a given read operation is uniquely paired off to some write or read operation iteration. Finally, channels that result from different read accesses from the same statement to data written by the same write access are combined into a single channel if this combination does not introduce reordering, a characteristic explained in the next section.

4.2. Determining channel types

In general, the channels we derived in the previous section may not be FIFOs. That is, data may be written to the channel in an order that is different from the order in which data is read. We therefore need to check whether such reordering occurs. This check can again be formulated as a (set of) PIP problem(s). Reordering occurs if and only if there exist two pairs of write and read iterations, $(\mathbf{w}_1, \mathbf{r}_1), (\mathbf{w}_2, \mathbf{r}_2) \in \mathbb{Z}^{n_1} \times \mathbb{Z}^{n_2}$, such that the order of the write operations is different from the order of the read operations, that is, $\mathbf{w}_1 \succ \mathbf{w}_2$ and $\mathbf{r}_1 \prec \mathbf{r}_2$, or equivalently

$$\mathbf{w}_1 - \mathbf{w}_2 \succ 0, \qquad \mathbf{r}_1 \prec \mathbf{r}_2. \quad (6)$$

Given a union of integer relations describing the channel (5), then for any pair of relations in this union, (D_{j_1}, D_{j_2}), we therefore need to solve n_2 PIP problems

$$\begin{aligned} \text{lexmax} \{ (\mathbf{t}, (\mathbf{w}_1, \mathbf{r}_1), (\mathbf{w}_2, \mathbf{r}_2), \mathbf{p}) \mid \\ (\mathbf{w}_1, \mathbf{r}_1) \in D_{j_1} \wedge (\mathbf{w}_2, \mathbf{r}_2) \in D_{j_2} \\ \wedge \mathbf{t} = \mathbf{w}_1 - \mathbf{w}_2 \wedge \mathbf{r}_1 \prec \mathbf{r}_2 \}, \end{aligned} \quad (7)$$

where $\mathbf{r}_1 \prec \mathbf{r}_2$ should be expanded according to (1) to obtain the n_2 problems. If any of these problems has a solution and if it is lexicographically positive or unbounded (in the first n_1 positions), then reordering occurs. Note that we do not compute the maximum of $\mathbf{t} = \mathbf{w}_1 - \mathbf{w}_2$ in terms of the parameters $\mathbf{p}$, but rather the maximum over all values of the parameters. If reordering occurs for any value of the parameters, then we simply consider the channel to be reordering. Equation (7) therefore actually represents a nonparametric

```
for (i = 0; i < N; ++i)
    a[i] = A(i);
for (j = 0; j < N; ++j)
    b[j] = B(j);
for (i = 0; i < N; ++i)
    for (j = 0; j < N; ++j)
        c[i][j] = a[i] * b[j];
```

FIGURE 2: Outer product source code.

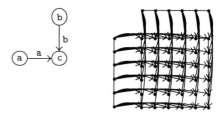

FIGURE 3: Outer product dependence graph with multiplicity.

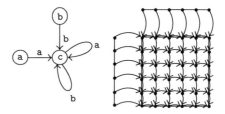

FIGURE 4: Outer product dependence graph without multiplicity.

integer programming problem. The large majority of these problems will be trivially unsatisfiable.

The reordering test of this section is a variation of the reordering test of [17], where it is formulated as $n_1 \times n_2$ PIP problems for a channel described by a single integer relation. A further difference is that the authors of [17] perform a more standard dataflow analysis and therefore also need to consider a second characteristic called *multiplicity*. Multiplicity occurs when the same data is read more than once from the same channel. Since we also consider previous reads from the same node as potential sources in our dataflow analysis, the channels we derive will never have multiplicity, but rather will be split into two channels, one corresponding to the first read and one self-loop channel propagating the value to subsequent reads.

Removing multiplicity not only reduces the communication between different nodes, it can also remove some artificial reorderings. A typical example of this situation is the outer product of two vectors, shown in Figure 2. Figure 3 shows the result of standard dataflow analysis. The left part of the figure shows the three nodes and two channels; the right part shows the data flow between the individual iterations of the nodes. The iterations are executed top-down, left-to-right. The channel between a and c is described by the relation

$$D_{a \to c} = \{(i_a, i_c, j_c) \mid 0 \le i_c \le N - 1 \\ \land\, 0 \le j_c \le N - 1 \land i_a = i_c\} \tag{8}$$

and would be classified as nonreordering, since the data elements are read (albeit multiple times) in the order in which they are produced. The channel between b and c, on the other hand, is described by the relation

$$D_{b \to c} = \{(j_b, i_c, j_c) \mid 0 \le i_c \le N - 1 \\ \land\, 0 \le j_c \le N - 1 \land j_b = j_c\} \tag{9}$$

and would be classified as reordering, with the further complication that the same data element needs to be sent over the channel multiple times. By simply letting node c only read a data element from these channels the first time it needs the data and from a newly introduced self-loop channel all other times, we obtain the network shown in Figure 4. In this network, all channels, including the new self-loop channels, are FIFOs. For example, the channel with dependence relation

$D_{b \to c}$ (9) is split into a channel with relation

$$D'_{b \to c} = \{(j_b, i_c, j_c) \mid i_c = 0 \land 0 \le j_c \le N - 1 \land j_b = j_c\} \tag{10}$$

and a self-loop channel with relation

$$D_{c \to c} = \{(i'_c, j'_c, i_c, j_c) \mid 1 \le i_c \le N - 1 \\ \land\, 0 \le j_c \le N - 1 \tag{11} \\ \land\, j'_c = j_c \land i'_c = i_c - 1\}.$$

4.3. Self-loops

When removing multiplicity from channels, our dataflow analysis introduces extra self-loop channels. Some of these channels can be further optimized. A simple, but important case is that where the channels hold at most one data element throughout the execution of the program. Such channels can be replaced by a single register. This situation occurs when for every pair of write and read iterations $(\mathbf{w}_2, \mathbf{r}_2)$, there is no other read iteration $\mathbf{r}_1$ reading from the same channel in between. In other words, the situation does not occur if and only if there exist two pairs of write and read iterations, $(\mathbf{w}_1, \mathbf{r}_1)$ and $(\mathbf{w}_2, \mathbf{r}_2)$, such that $\mathbf{w}_2 \prec \mathbf{r}_1 \prec \mathbf{r}_2$, or equivalently $\mathbf{r}_1 - \mathbf{w}_2 \succ \mathbf{0}$ and $\mathbf{r}_1 \prec \mathbf{r}_2$. Notice the similarity between this condition and the reordering condition (6). The PIP problems that need to be solved to determine this condition are therefore nearly identical to the problems (7), namely.,

$$\text{lexmax}\, \{(\mathbf{t}, (\mathbf{w}_1, \mathbf{r}_1), (\mathbf{w}_2, \mathbf{r}_2), \mathbf{p}) \mid \\ (\mathbf{w}_1, \mathbf{r}_1) \in D_{j_1} \land (\mathbf{w}_2, \mathbf{r}_2) \in D_{j_2} \tag{12} \\ \land\, \mathbf{t} = \mathbf{r}_1 - \mathbf{w}_2 \land \mathbf{r}_1 \prec \mathbf{r}_2\},$$

where again (D_{j_1}, D_{j_2}) is a pair of relations in the union describing the channel and where $\mathbf{r}_1 \prec \mathbf{r}_2$ should be expanded according to (1).

If such a channel has the additional property that the single value it contains is always propagated to the next iteration of the node (a condition that can again be checked using PIP), then we remove the channel completely and attach the register to the input argument of the function and call the FIFO(s) that read the value for the first time "sticky FIFOs." This is a special case of the optimization applied to in-order channels with multiplicity of [17] that allows for slightly more efficient implementations due to the extra property.

Another special case occurs when the number of iterations of the node between a write to the self-loop channel and the corresponding read is a constant, which we can determine by simply counting the number of intermediate iterations (symbolically) and checking whether the result is a constant function. In this case, we can replace the FIFO by a shift register, which can be implemented more efficiently in hardware. Note, however, that there may be a trade-off since the size of the channel as a shift register (i.e., the constant function above) may be larger than the size of the channel as a FIFO. On the other hand, the FIFO size may be more difficult to determine (see Section 5.2).

5. COMPUTING CHANNEL SIZES

In this section, we explain how we compute the buffer sizes for the FIFOs in our networks at compile-time. This computation may not be feasible for process networks in general, but we are dealing here with the easier case of networks generated from static affine nested loop programs. We first consider self-loops, with a special case in Section 5.1, and the general case in Section 5.2. In Section 5.3, we then explain how to reduce the general case of FIFOs to self-loops by scheduling the network.

5.1. Uniform self-dependences on rectangular domains

An important special case occurs when the channel is represented by a single integer relation that in turn represents a uniform dependence over a rectangular domain. A dependence is called uniform if the difference between the read and write iteration vectors is a (possibly parametric) constant over the whole relation. We call such a dependence a uniform dependence over a rectangular domain if the set of iterations reading from the channel form a rectangular domain. (Note that due to the dependence being uniform, also the write iterations will form a rectangular domain in this case.) For example, the relation $D_{c \to c}$ (11) from Section 4.2 is a uniform dependence over a rectangular domain since the difference between the read and write iteration vectors is $(i_c, j_c) - (i'_c, j'_c) = (1, 0)$ and since the projection onto the read iterations is the rectangle $1 \le i_c \le N - 1 \wedge 0 \le j_c \le N - 1$.

The required buffer size is easily calculated in these cases since in each (overlapping) iteration of any of the loops in the loop nest, the number of data elements produced is exactly the same as the number of elements consumed. The channel will therefore never contain more data elements than right

before the first data element is read, or equivalently, right after the last data element is written. To compute the buffer size, we therefore simply need to take the first read iteration and count the number of write iterations that are lexicographically smaller than this read iteration using `barvinok`. In the example, the first read operation occurs at iteration $(1, 0)$ and so we need to compute

$$\#(S \cap \{(i'_c, j'_c) \mid i'_c < 1\}) + \#(S \cap \{(i'_c, j'_c) \mid i'_c = 1 \wedge j'_c < 0\}),$$
(13)

with S the set of write iterations

$$S = \{(i'_c, j'_c) \mid 0 \le i'_c \le N - 2 \wedge 0 \le j'_c \le N - 1\}. \quad (14)$$

The result of this computation is $N + 0 = N$.

5.2. General self-loop FIFOs

An easy approximation can be obtained by computing the number of array elements in the original program that are written to the channel. That is, we can intersect the domain of write iterations with the access relation and project onto the array space. The resulting (union of) sets can be enumerated symbolically using `barvinok`. The result may however be a large overestimate of the actual buffer size requirements.

The actual amount of data in a channel at any given iteration can be computed fairly easily. We simply compute the number of read iterations that are executed before a given read operation and subtract the resulting expression from the number of write iterations that are executed before the given read operation. This computation can again be performed entirely symbolically and the result is a piecewise (quasi-)polynomial in the read iterators and the parameters. The required buffer size is the maximum of this expression over all read iterations.

For sufficiently regular problems, we can compute the above maximum symbolically by performing some simplifications and identifying some special cases. In the general case, we can apply Bernstein expansion [25] to obtain a parametric *upper bound* on the expression. For *nonparametric* problems, however, it is usually easier to *simulate* the communication channel. That is, we use CLooG [27] to generate code that increments a counter for each iteration writing to the channel and decrements the counter for each read iteration. The maximum value attained by this counter is recorded and reflects the channel size.

5.3. Nonself-loop FIFOs

Computing the sizes of self-loop channels is relatively easy because the order of execution within a node of the network is fixed. However, the relative order of iterations from different nodes is not known a priori since this order is determined at run-time. Computing minimal deadlock-free buffer sizes is a nontrivial global optimization problem. This problem becomes easier if we first compute a deadlock-free schedule and then compute the buffer sizes for each channel individually. Note that this schedule is only computed for the purpose

```
for (j=0; j < Nrw; j++)
  for (i=0; i < Ncl; i++)
    a[j][i] = ReadImage();

for (j=1; j < Nrw-1; j++)
  for (i=1; i < Ncl-1; i++)
    Sbl[j][i] = Sobel(a[j-1][i-1], a[j][i-1], a[j+1][i-1],
                      a[j-1][ i], a[j][ i], a[j+1][ i],
                      a[j-1][i+1], a[j][i+1], a[j+1][i+1]);
```

FIGURE 5: Source code of a Sobel edge detection example.

of computing the buffer sizes and is discarded afterward. The schedule we compute may not be optimal and the resulting buffer sizes may not be valid for the optimal schedule. Our computations do ensure, however, that a valid schedule exists for the computed buffer sizes.

The schedule is computed using a greedy approach. This approach may not work for process networks in general, but it does work for any network derived from an SANLP. The basic idea is to place all iteration domains in a common iteration space at an offset that is computed by the scheduling algorithm. As in the individual iteration spaces, the execution order in this common iteration space is the lexicographical order. By fixing the offsets of the iteration domain in the common space, we have therefore fixed the relative order between any pair of iterations from any pair of iteration domains. The algorithm starts by computing for any pair of connected nodes, the minimal dependence distance vector, a distance vector being the difference between a read operation and the corresponding write operation. Then the nodes are greedily combined, ensuring that all minimal distance vectors are (lexicographically) positive. The end result is a schedule that ensures that every data element is written before it is read. For more information on this algorithm, we refer to [28], where it is applied to perform loop fusion on SANLPs. Note that unlike the case of loop fusion, we can ignore antidependences here, unless we want to use the declared size of an array as an estimate for the buffer size of the corresponding channels. (Antidependences are ordering constraints between reads and subsequent writes that ensure an array element is not overwritten before it is read.)

After the scheduling, we may consider all channels to be self-loops of the common iteration space and we can apply the techniques from the previous sections with the following qualifications. We will not be able to compute the absolute minimum buffer sizes, but at best the minimum buffer sizes for the computed schedule. We cannot use the declared size of an array as an estimate for the channel size, unless we have taken into account antidependences. An estimate that remains valid is the number of write iterations.

We have tacitly assumed above that all iteration domains have the same dimension. If this is not the case, then we first need to assign a dimension of the common (bigger) iteration space to each of the dimensions of the iteration domains of lower dimension. For example, the single iterator of the first loop of the program in Figure 2 would correspond to the outer loop of the 2D common iteration space, whereas the single iterator of the second loop would correspond to the inner loop, as shown in Figure 3. We currently use a greedy heuristic to match these dimensions, starting from domains with higher dimensions and matching dimensions that are related through one or more dependence relations. During this matching we also, again greedily, take care of any scaling that may need to be performed to align the iteration domains. Although our heuristics seem to perform relatively well on our examples, it is clear that we need a more general approach such as the linear transformation algorithm of [29].

6. WORKED-OUT EXAMPLES

In this section, we show the results of applying our optimization techniques to two image processing algorithms. The generated process networks (PN) enjoy a reduction in the amount of data transferred between nodes and reduced memory requirements, resulting in a better performance, that is, a reduced execution time. The first algorithm is the Sobel operator, which estimates the gradient of a 2D image. This algorithm is used for edge detection in the preprocessing stage of computer vision systems. The second algorithm is a forward discrete wavelet transform (DWT). The wavelet transform is a function for multiscale analysis and has been used for compact signal and image representations in denoising, compression, and feature detection processing problems for about twenty years.

6.1. Sobel edge detection

The Sobel edge detection algorithm is described by the source code in Figure 5. To estimate the gradient of an image, the algorithm performs a convolution between the image and a 3×3 convolution mask. The mask is slid over the image, manipulating a square of 9 pixels at a time, that is, each time 9 image pixels are read and 1 value is produced. The value represents the approximated gradient in the center of the processed image area. Applying the regular dataflow analysis on this example using Compaan results in the process network (PN)

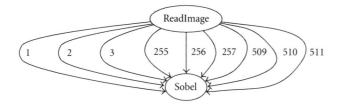

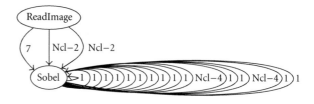

FIGURE 6: Compaan generated process network for the Sobel example.

FIGURE 7: The generated process network for the Sobel example using the self reuse technique.

depicted in Figure 6. It contains 2 nodes (representing the `ReadImage` and `Sobel` functions) and 9 channels (representing the arguments of the `Sobel` function). Each channel is marked with a number showing the buffer size it requires. These numbers were obtained by running a simulation processing an image of 256×256 pixels (`Nrw=Ncl=256`). The `ReadImage` node reads the input image from memory pixel by pixel and sends it to the `Sobel` node through the 9 channels. Since the 9 pixel values are read in parallel, the executions of the `Sobel` node can start after reading 2 lines and 3 pixels from memory.

After detecting self reuse through read accesses from the same statement as described in Section 4.1, we obtain the PN in Figure 7. Again, the numbers next to each channel specify the buffer sizes of the channels. We calculated them at compile time using the techniques described in Section 5. The number of channels between the nodes is reduced from 9 to 3 while several self-loops are introduced. Reducing the communication load between nodes is an important issue since it influences the overall performance of the final implementation. Each data element transferred between two nodes introduces a communication overhead which depends on the architecture of the system executing the PN. For example, if a PN is mapped onto a multiprocessor system with a shared bus architecture, then the 9 pixel values are transferred sequentially through the shared bus, even though in the PN model they are specified as 9 (parallel) channels (Figure 6). In this example it is clear that the PN in Figure 7 will only suffer a third of the communication overhead because it contains 3 times fewer channels between the nodes. The self-loops are implemented using the local processor memory and they do not use the communication resources of the system. Moreover, most of the self-loops require only 1 register which makes their implementations simpler than the implementation of a communication channel (FIFO). This also holds for PNs implemented as dedicated hardware. A single-register self-loop is much cheaper to implement in terms of HW resources than a FIFO channel. Another important issue (in both SW and HW systems) is the memory requirement. For the PN in Figure 6 the total amount of memory required is 2304 locations, while the PN in Figure 7 requires only 1033 (for a 256×256 image). This shows that the detection of self reuse reduces the memory requirements by a factor of more than 2.

In principle, the three remaining channels between the two nodes could be combined into a single channel, but, due

to boundary conditions, the order in which data would be read from this channel is different from the order in which it is written and we would therefore have a reordering channel (see Section 4.2). Since the implementation of a reordering channel is much more expensive than that of a FIFO channel, we do not want to introduce such reordering. The reason we still have 9 channels (7 of which are combined into a single channel) after reuse detection is that each access reads at least some data for the first time. We can change this behavior by extending the loops with a few iterations, while still only reading the same data as in the original program. All data will then be read for the first time by access `a[j+1][i+1]` only, resulting in a single FIFO between the two nodes. To ensure that we only read the required data, some of the extra iterations of the accesses do not read any data. We can (manually) effectuate this change in C by using (implicit) temporary variables and, depending on the index expressions, reading from "`noise`," as shown in Figure 8. By using the simple copy propagation technique of Section 3.2, these modifications do not increase the number of nodes in the PN.

The generated optimized PN shown in Figure 9 contains only one (FIFO) channel between the `ReadImage` and `Sobel` nodes. All other communications are through self-loops. Thus, the communication between the nodes is reduced 9 times compared to the initial PN (Figure 6). The total memory requirements for a 256×256 image have been reduced by a factor of almost 4.5 to 519 locations. Note that the results of the extra iterations of the `Sobel` node, which partly operate on "noise," are discarded and so the final behavior of the algorithm remains unaltered. However, with the reduced number of communication channels and overhead, the final (real) implementation of the optimized PN will have a better performance.

6.2. Discrete wavelet transform

In the discrete wavelet transform (DWT) the input image is decomposed into different decomposition levels. These decomposition levels contain a number of subbands, which consist of coefficients that describe the horizontal and vertical spatial frequency characteristics of the original image. The DWT requires the signal to be extended periodically. This periodic symmetric extension is used to ensure that for the filtering operations that take place at both boundaries of the signal, one signal sample exists and spatially

```
#define A(j,i) (j>=0 && i>=0 && i<Ncl ? a[j][i] : noise)
#define S(j,i) (j>=1 && i>=1 && i<Ncl-1 ? Sbl[j][i] : noise)

for (j=0; j < Nrw; j++)
  for (i=0; i < Ncl; i++)
    a[j][i] = ReadImage();

for (j=-1; j < Nrw-1; j++)
  for (i=-1; i < Ncl+1; i++)
    S(j,i) = Sobel(A(j-1, i-1), A(j, i-1), A(j+1, i-1),
                   A(j-1,   i), A(j,   i), A(j+1,   i),
                   A(j-1, i+1), A(j, i+1), A(j+1, i+1));
```

FIGURE 8: Modified source code of the Sobel edge detection example.

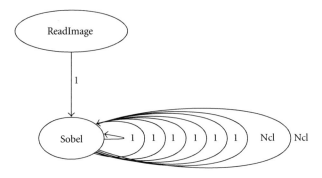

FIGURE 9: The generated PN for the modified Sobel edge detection example.

corresponds to each coefficient of the filter mask. The number of additional samples required at the boundaries of the signal is therefore filter-length dependent.

The C program realizing one level of a 2D forward DWT is presented in Figure 10. In this example, we use a lifting scheme of a reversible transformation with 5/3 filter [30]. In this case the image has to be extended with one pixel at the boundaries. All the boundary conditions are described by the conditions in code lines 8, 11, 17, 20, 26, and 29.

First, a 1D DWT is applied in the vertical direction (lines 7 to 13). Two intermediate variables are produced (low- and high-pass filtered images subsampled by 2—lines 9 and 12). They are further processed by a 1D DWT applied in the horizontal direction and thus producing (again subsampled by 2) a four subbands decomposition: HL (line 18), LL (line 21), HH (line 27), and LH (line 30). The process network generated by using the regular dataflow analysis (and Compaan tool) is depicted in Figure 11. The PN contains 23 nodes, half of them just copying pixels at the boundaries of the image. Channel sizes are estimated by running a simulation again processing an image 256×256 pixels. Although most of the channels have size 1, they cannot be implemented by a simple register since they connect nodes and additional logic (FIFO like) is required for synchronization. Obviously, the generated PN has considerable initial overhead.

The optimization goals for this example are to remove the Copy nodes and to reduce the communication between the nodes as much as possible. We achieve these goals by applying our techniques. The optimized process network is shown in Figure 12. The simple copy propagation technique reduces the number of the nodes from 23 to 11 and the detection of self reuse technique reduces the communication between the nodes from 40 to 15 channels introducing 8 self-loop channels. There is only one channel connecting two nodes of the PN in Figure 12, except for the channels between the ReadImage and high_filt_vert nodes. In this case, we detect that a combined channel would be reordering. As we mentioned in the previous example, we prefer not to introduce reordering and therefore generate more (FIFO) channels. As a result, the number of channels emanating from the ReadImage has been reduced by only one compared to the initial PN (Figure 11). The buffer sizes are calculated at compile time using our techniques described in Section 5 and the correctness of the process network is tested using the YAPI environment [5]. Note that in this example applying the optimization techniques has little effect on the memory requirements: the number of memory locations required for an image of 256×256 pixels is 2585 compared to 2603 for the initial DWT PN. However, the topology of the optimized PN has been simplified significantly allowing an efficient HW and/or SW implementation.

7. COMPARISON TO COMPAAN AND COMPAAN-LIKE NETWORKS

Table 1 compares the number of channels in Compaan-like networks to the number of channels in our networks. The Compaan-like networks were generated by using standard dataflow analysis instead of also considering reads as sources and by turning off the copy propagation of temporary scalars and the combination of channels reading from the same write access. The table shows a decomposition of the channels into different types. In-Order (IO) and Out-of-Order (OO) refer to FIFOs and reordering channels, respectively, and the M-suffix refers to multiplicity, which does not occur in our networks. Each column is further split into

TABLE 1: Comparison to channel numbers of Compaan-like networks.

Algorithm name	Compaan-like networks				Our networks	
	IO	IOM	OO	OOM	IO	OO
	sl + ed	sl + ed	sl + ed	sl + ed	1r + sl + ed	sl + ed
LU-Factor	3 + 13	1 + 7	0 + 3	0 + 1	2 + 5 + 16	0 + 3
QR-Decomp	4 + 8	0 + 0	0 + 0	0 + 0	1 + 3 + 8	0 + 0
SVD	4 + 41	0 + 4	0 + 18	0 + 0	8 + 0 + 34	0 + 16
Faddeev	3 + 20	0 + 3	0 + 1	0 + 0	4 + 2 + 19	0 + 1
Gauss-Elim.	2 + 5	0 + 0	0 + 1	1 + 2	0 + 6 + 6	0 + 1
Motion Est.	27 + 66	0 + 0	0 + 0	0 + 0	0 + 54 + 66	0 + 0
M-JPEG	9 + 21	0 + 17	0 + 0	0 + 0	18 + 0 + 38	0 + 0

```
      for (i = 0; i < 2*Nrw; i++)
        for (j = 0; j < 2*Ncl; j++)
          a[i][j] = ReadImage();

5     for (i = 0; i < Nrw; i++) {
        // 1D DWT in vertical direction with subsampling
        for (j = 0; j < 2*Ncl; j++) {
          tmpLine = (i==Nrw-1) ? a[2*i][j] : a[2*i+2][j];
          Hf[j] = high_flt_vert(a[2*i][j], a[2*i+1][j], tmpLine);
10
          tmp = (i==0) ? Hf[j] : oldHf[j];
          low_flt_vert(tmp, a[2*i][j], Hf[j], &oldHf[j], &Lf[j]);
        }

15      // 1D DWT in horizontal direction with subsampling ---------
        for (j = 0; j < Ncl; j++) {
          tmp = (j==Ncl-1) ? Lf[2*j] : Lf[2*j+2];
          HL[i][j] = high_flt_hor(Lf[2*j], Lf[2*j+1], tmp);

20        tmp = (j==0) ? HL[i][j] : HL[i][j-1];
          LL[i][j] = low_flt_hor(tmp, Lf[2*j],  HL[i][j]);
        }

        // 1D DWT in horizontal direction with subsampling ---------
25      for (j = 0; j < Ncl; j++) {
          tmp = (j==Ncl-1) ? Hf[2*j] : Hf[2*j+2];
          HH[i][j] = high_flt_hor(Hf[2*j], Hf[2*j+1], tmp);

          tmp = (j == 0) ? HH[i][j] : HH[i][j-1];
30        LH[i][j] = low_flt_hor(tmp, Hf[2*j], HH[i][j]);
        }
      }
      // The Outputs -------------------------------------------------
      for (i = 0; i < Nrw; i++)
35      for (j = 0; j < Ncl; j++) {
          Sink(LL[i][j]);
          Sink(HL[i][j]);
          Sink(LH[i][j]);
          Sink(HH[i][j]);
40      }
```

FIGURE 10: Source code of a discrete wavelet transform example.

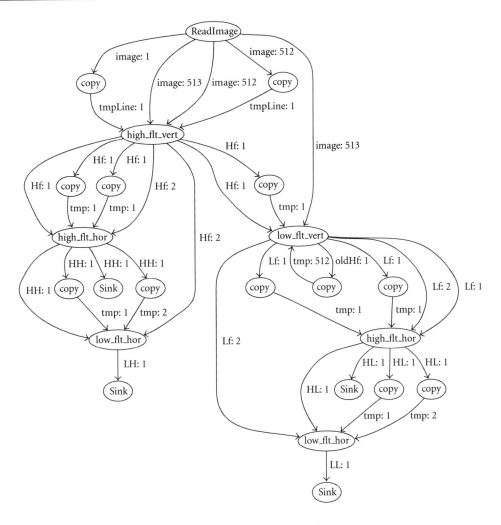

FIGURE 11: 2D-DWT process network with copy nodes.

self-loops+edges, or single-register+self-loops+edges for our FIFOs.

Note that our numbers on Compaan-like networks differ from those on Compaan networks reported in [17]. Due to a difference in internal representation, some of our channels are split into several Compaan-channels. In Compaan, these channels are recombined, with possibly further combinations, at a later stage. From the table, we can conclude that our techniques have split all OOM channels in examples LU-Factor and Gauss-Elim. into pairs of FIFOs. In general, we also have fewer channels between different nodes at the expense of more self-loops, which are a lot more efficient. For example, for SVD, the number of edges is reduced from 63 to 50, while for LU-Factor we have a reduction from 24 to 19 and for Faddeev from 24 to 20. Finally, we are able to identify (in examples LU-Factor, QR-Decomp, SVD, Faddeev, and M-JPEG) that many of these self-loops are "single-register" FIFOs, where "register" should be interpreted as "token," which may be a whole table in the case of M-JPEG.

As to the time needed to derive the networks, Compaan itself takes 2.3 to 28.1 seconds on the examples in Table 1,

while our tool takes 2.5 to 46.4 seconds. Most of the latter time is spent in the computation of the FIFO sizes, which Compaan does not compute.

8. CONCLUSIONS AND DISCUSSION

In this paper, we have improved upon the state-of-the-art conversion of sequential programs to process networks in several ways. We have shown that we can reduce the number of reordering channels as well as the total number of channels between different nodes by extending the standard dataflow analysis to detect reuse within a node. This effect is enhanced by first removing the (artificial) copy nodes introduced by Compaan through simple copy propagation. Our modified dataflow analysis leads to a removal of all reordering channels with multiplicity that appear in our examples and a reduction of the communication volume by up to a factor 9 in the extreme case. We have further shown how to compute the FIFO sizes exactly for self-loops in nonparametric programs and approximately for other channels and self-loops in parametric programs.

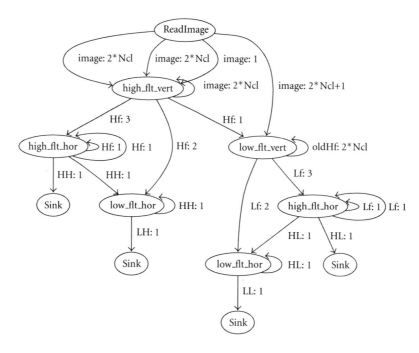

Figure 12: Optimized 2D-DWT process network.

ACKNOWLEDGMENTS

This research is partly supported by PROGRESS, the Embedded Systems and Software Research Program of the Dutch Technology Foundation STW—Project ARTEMISIA (LES.6389). We also thank Bart Kienhuis for his help and for the discussions on some of the topics in this paper.

REFERENCES

[1] A. Darte, R. Schreiber, and G. Villard, "Lattice-based memory allocation," *IEEE Transactions on Computers*, vol. 54, pp. 1242–1257, 2005.

[2] E. A. Lee and A. Sangiovanni-Vincentelli, "A framework for comparing models of computation," *IEEE Transactions on Computer-Aided Design of Integrated Circuits and Systems*, vol. 17, no. 12, pp. 1217–1229, 1998.

[3] J. Davis, R. Galicia, M. Goel, et al., "PtolemyII: heterogeneous concurrent modeling and design in java," Tech. Rep. UCB/ERL M99/40, University of California, Berkeley, Calif, USA, 1999.

[4] G. Kahn, "The semantics of a simple language for parallel programming," in *Proceedings of the IFIP Congress*, pp. 471–475, North-Holland, Stockholm, Sweden, August 1974.

[5] E. A. de Kock, G. Essink, W. J. M. Smits, et al., "YAPI: application modeling for signal processing systems," in *Proceedings of the 37th Design Automation Conference (DAC '00)*, pp. 402–405, Los Angeles, Calif, USA, June 2000.

[6] E. A. de Kock, "Multiprocessor mapping of process networks: a JPEG decoding case study," in *Proceedings of the 15th International Symposium on System Synthesis (ISSS '02)*, pp. 68–73, Kyoto, Japan, October 2002.

[7] B. K. Dwivedi, A. Kumar, and M. Balakrishnan, "Automatic synthesis of system on chip multiprocessor architectures for process networks," in *Proceedings of the 2nd IEEE/ACM/IFIP International Conference on Hardware/Software Codesign and Systems Synthesis, (CODES+ISSS '04)*, pp. 60–65, IEEE Computer Society, Stockholm, Sweden, September 2004.

[8] K. Goossens, J. Dielissen, J. van Meerbergen, et al., "Guaranteeing the quality of services in networks on chip," in *Networks on Chip*, pp. 61–82, Kluwer Academic Publishers, Hingham, Mass, USA, 2003.

[9] P. Lieverse, T. Stefanov, P. van der Wolf, and E. Deprettere, "System level design with SPADE: an M-JPEG case study," in *Proceedings of the International Conference on Computer-Aided Design (ICCAD '01)*, pp. 31–38, San Jose, Calif, USA, November 2001.

[10] A. Nieuwland, J. Kang, O. P. Gangwal, et al., *C-HEAP: A Heterogeneous Multi-Processor Architecture Template and Scalable and Flexible Protocol for the Design of Embedded Signal Processing Systems*, Kluwer Academic Publishers, Norwell, Mass, USA, 2002.

[11] H. Nikolov, T. Stefanov, and E. Deprettere, "Multi-processor system design with ESPAM," in *Proceedings of the 4th IEEE/ACM/IFIP International Conference on Hardware/Software Codesign and System Synthesis (CODES+ISSS '06)*, pp. 211–216, Seoul, Korea, October 2006.

[12] A. D. Pimentel, C. Erbas, and S. Polstra, "A systematic approach to exploring embedded system architectures at multiple abstraction levels," *IEEE Transactions on Computers*, vol. 55, no. 2, pp. 99–112, 2006.

[13] T. Stefanov, C. Zissulescu, A. Turjan, B. Kienhuis, and E. Deprettere, "System design using Kahn process networks: the Compaan/Laura approach," in *Proceedings of Conference Design, Automation and Test in Europe (DATE '04)*, vol. 1, pp. 340–345, Paris, France, February 2004.

[14] P. van der Wolf, P. Lieverse, M. Goel, D. La Hei, and K. Vissers, "MPEG-2 decoder case study as a driver for a system level

design methodology," in *Proceedings of the 7th International Workshop on Hardware/Software Codesign (CODES '99)*, pp. 33–37, Rome, Italy, May 1999.

[15] B. Kienhuis, E. Rijpkema, and E. Deprettere, "Compaan: deriving process networks from matlab for embedded signal processing architectures," in *Proceedings of the 8th International Workshop Hardware/Software Codesign (CODES '00)*, pp. 13–17, ACM Press, San Diego, Calif, USA, May 2000.

[16] E. Rijpkema, E. F. Deprettere, and B. Kienhuis, "Deriving process networks from nested loop algorithms," *Parallel Processing Letters*, vol. 10, no. 2, pp. 165–176, 2000.

[17] A. Turjan, B. Kienhuis, and E. Deprettere, "Translating affine nested-loop programs to process networks," in *Proceedings of International Conference on Compilers, Architecture, and Synthesis for Embedded Systems (CASES '04)*, pp. 220–229, Washington, DC, USA, September 2004.

[18] A. Darte, R. Schreiber, and G. Villard, "Lattice-based memory allocation," in *Proceedings of the International Conference on Compilers, Architecture, and Synthesis for Embedded Systems (CASES '03)*, pp. 298–308, ACM Press, San Jose, Calif, USA, October-November 2003.

[19] K. Beyls and E. H. D'Hollander, "Generating cache hints for improved program efficiency," *Journal of Systems Architecture*, vol. 51, no. 4, pp. 223–250, 2005.

[20] T. Vander Aa, M. Jayapala, F. Barat, H. Corporaal, F. Catthoor, and G. Deconinck, "A high-level memory energy estimator based on reuse distance," in *Proceedings of the 3rd Workshop on Optimizations for DSP and Embedded Systems (ODES '05)*, San Jose, Calif, USA, March 2005.

[21] P. Vanbroekhoven, G. Janssens, M. Bruynooghe, H. Corporaal, and F. Catthoor, "Advanced copy propagation for arrays," in *Proceedings of the ACM SIGPLAN Conference on Languages, Compilers, and Tools for Embedded Systems (LCTES '03)*, U. Kremer, Ed., pp. 24–33, ACM Press, San Diego, Calif, USA, June 2003.

[22] P. Feautrier, "Automatic parallelization in the polytope model," in *The Data Parallel Programming Model*, vol. 1132 of *Lecture Notes in Computer Science*, pp. 79–103, Springer, London, UK, 1996.

[23] P. Feautrier, "Parametric integer programming," *Operationnelle/Operations Research*, vol. 22, no. 3, pp. 243–268, 1988.

[24] S. Verdoolaege, R. Seghir, K. Beyls, V. Loechner, and M. Bruynooghe, "Analytical computation of Ehrhart polynomials: enabling more compiler analyses and optimizations," in *Proceedings of the International Conference on Compilers, Architecture, and Synthesis for Embedded Systems (CASES '04)*, pp. 248–258, Washington, DC, USA, September 2004.

[25] P. Clauss, F. J. Fernández, D. Gabervetsky, and S. Verdoolaege, "Symbolic polynomial maximization over convex sets and its application to memory requirement estimation," ICPS Research Reports 06–04, Université Louis Pasteur, Strasbourg, France, 2006, http://icps.u-strasbg.fr/upload/icps-2006-173.pdf.

[26] P. Feautrier, "Dataflow analysis of array and scalar references," *International Journal of Parallel Programming*, vol. 20, no. 1, pp. 23–53, 1991.

[27] C. Bastoul, "Code generation in the polyhedral model is easier than you think," in *Proceedings of the 13th International Conference on Parallel Architectures and Compilation Techniques (PACT '04)*, pp. 7–16, IEEE Computer Society, Antibes Juan-les-Pins, France, September 2004.

[28] S. Verdoolaege, M. Bruynooghe, G. Janssens, and F. Catthoor, "Multi-dimensional incremental loop fusion for data locality," in *Proceedings of the 14th IEEE International Conference on Application-Specific Systems, Architectures, and Processors (ASAP '03)*, D. Martin, Ed., pp. 17–27, The Hague, The Netherlands, June 2003.

[29] S. Verdoolaege, K. Danckaert, F. Catthoor, M. Bruynooghe, and G. Janssens, "An access regularity criterion and regularity improvement heuristics for data transfer optimization by global loop transformations," in *Proceedings of the 1st Workshop on Optimization for DSP and Embedded Systems (ODES '03)*, San Francisco, Calif, USA, March 2003.

[30] I. Daubechies and W. Sweldens, "Factoring wavelet transforms into lifting steps," *Journal of Fourier Analysis and Applications*, vol. 4, no. 3, pp. 247–269, 1998.

Hindawi Publishing Corporation
EURASIP Journal on Embedded Systems
Volume 2007, Article ID 47580, 22 pages
doi:10.1155/2007/47580

Research Article

A SystemC-Based Design Methodology for Digital Signal Processing Systems

Christian Haubelt, Joachim Falk, Joachim Keinert, Thomas Schlichter, Martin Streubühr, Andreas Deyhle, Andreas Hadert, and Jürgen Teich

Hardware-Software-Co-Design, Department of Copmuter Sciences, Friedrich-Alexander-University of Erlangen-Nuremberg, 91054 Erlangen, Germany

Received 7 July 2006; Revised 14 December 2006; Accepted 10 January 2007

Recommended by Shuvra Bhattacharyya

Digital signal processing algorithms are of big importance in many embedded systems. Due to complexity reasons and due to the restrictions imposed on the implementations, new design methodologies are needed. In this paper, we present a SystemC-based solution supporting *automatic design space exploration, automatic performance evaluation*, as well as *automatic system generation* for mixed hardware/software solutions mapped onto FPGA-based platforms. Our proposed hardware/software codesign approach is based on a SystemC-based library called SysteMoC that permits the expression of different models of computation well known in the domain of digital signal processing. It combines the advantages of executability and analyzability of many important models of computation that can be expressed in SysteMoC. We will use the example of an MPEG-4 decoder throughout this paper to introduce our novel methodology. Results from a five-dimensional design space exploration and from automatically mapping parts of the MPEG-4 decoder onto a Xilinx FPGA platform will demonstrate the effectiveness of our approach.

1. INTRODUCTION

Digital signal processing algorithms, as for example real-time image enhancement, scene interpretation, or audio and video coding, have gained enormous popularity in embedded system design. They encompass a large variety of different algorithms, starting from simple linear filtering up to entropy encoding or scene interpretation based on neuronal networks. Their implementation, however, is very laborious and time consuming, because many different and often conflicting criteria must be met, as for example high throughput and low power consumption. Due to this rising complexity of these digital signal processing applications, there is demand for new design automation tools at a high level of abstraction.

Many design methodologies are proposed in the literature for exploring the design space of implementations of digital signal processing algorithms (cf. [1, 2]), but none of them is able to fully automate the design process. In this paper, we will close this gap by proposing a novel approach based on SystemC [3–5], a C++ class library, and state-of-the-art design methodologies. The proposed approach permits the design of digital signal processing applications with minimal designer interaction. The major advantage with respect to existing approaches is the combination of executability of the specification, exploration of implementation alternatives, and the usability of formal analysis techniques for restricted models of computation. This is achieved through restricting SystemC such that we are able to automatically detect the underlying model of computation (MoC) [6]. Our design methodology comprises the *automatic design space exploration* using state-of-the-art multiobjective evolutionary algorithms, the *performance evaluation* by automatically generating efficient simulation models, and *automatic platform-based system generation*. The overall design flow as proposed in this paper is shown in Figure 1 and is currently implemented in the framework SystemCoDesigner.

Starting with an executable specification written in SystemC, the designer can specify the target architecture template as well as the mapping constraints of the SystemC modules. In order to automate the design process, the SystemC application has to be written in a synthesizable subset of SystemC, called SysteMoC [7], and the target architecture template must be built from components supported by our component library. The components in the component

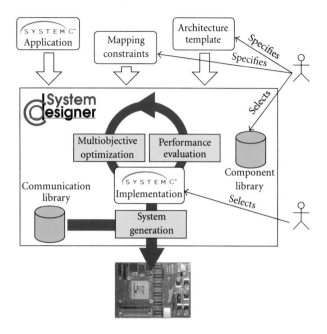

FIGURE 1: SystemCoDesigner design flow: for a given executable specification written in SystemC, the designer has to specify the architecture template as well as mapping constraints. The design space exploration is performed automatically using multiobjective evolutionary algorithms and is guided by an automatic simulation-based performance evaluation. Finally, any selected implementation can be automatically mapped efficiently onto an FPGA-based platform.

library are either written by hand using a hardware description language or can be taken from third party vendors. In this work, we will use IP cores especially provided by Xilinx. Furthermore, it is also possible to synthesize SysteMoC actors to RTL Verilog or VHDL using high-level synthesis tools as Mentor CatapultC [8] or Forte Cynthesizer [9]. However, there are limitations imposed on the actors given by these tools. As this is beyond the scope of this paper, we will omit discussing these issues here.

With this specification, the SystemCoDesigner design process is automated as much as possible. Inside SystemCo-Designer, a multiobjective evolutionary optimization (MO-EA) strategy is used in order to perform design space exploration. The exploration is guided by a simulation-based performance evaluation. Using SysteMoC as a specification language for the application, the generation of the simulation model inside the exploration can be automated. Then, the designer can carry out the decision making and select a design point for implementation. Finally, the platform-based implementation is generated automatically.

The remainder of this paper is dedicated to the different issues arising during our proposed design flow. Section 3 discusses the input format based on SystemC called SysteMoC. SysteMoC is a library based on SystemC that allows to describe and simulate communicating actors. The particularity of this library for actor-based design is to separate *actor functionality* and *communication behavior*. In particular, the separation of actor firing rules and communication behavior

is achieved by an explicit finite state machine model associated with each actor. This finite state machine permits the identification of the underlying model of computation of the SystemC application and, hence, if possible, allows to analyze the specification with formal techniques for properties such as boundedness of memory, (periodic) schedulability, deadlocks, and so forth.

Section 4 presents the model and the tasks performed during design space exploration. As the SysteMoC description only models the specified behavior of our system, we need additional information in order to perform *system-level synthesis*. Following the *Y-chart approach* [10, 11], a formal model of architecture (MoA) must be specified by the designer as well as mapping constraints for the actors in the SysteMoC description. With this formal model the system-level synthesis task is twofold: (1) determine the *allocation* of resources from the architecture template and (2) determine a *binding* of SystemC modules (actors) onto the allocated resources. During design space exploration, many implementations are constructed by the system-level exploration tool SystemCoDesigner. Each resulting implementation must be evaluated regarding different properties such as area, power consumption, performance, and so forth. Especially the performance evaluation, that is, latency and throughput, is critical in the context of digital signal processing applications. In our proposed methodology, we will use, beside others, a *simulation-based* approach. We will show how SysteMoC might help to automatically generate efficient simulation models during exploration.

In Section 5 our approach to automatic platform-based system synthesis will be presented targeting in our examples a Xilinx Virtex-II Pro FPGA-based platform. The key idea is to *generate a platform*, perform *software synthesis*, and provide *efficient communication channels* for the implementation. The results obtained by the synthesis will be compared to the simulation models generated during a five-dimensional design space exploration in Section 6. We will use the example of an MPEG-4 decoder throughout this paper to present our methodology.

2. RELATED WORK

In this section, we discuss some tools which are available for the design and synthesis of digital signal processing algorithms onto mixed and possibly multicore system-on-a-chip (SoC). Sesame (simulation of embedded system architectures for multilevel exploration) [12] is a tool for performance evaluation and exploration of heterogeneous architectures for the multimedia application domain. The applications are given by Kahn process networks modeled with a C++ class library. The architecture is modeled by architecture building blocks taken from a library. Using a SystemC-based simulator at transaction level, performance evaluation can be done for a given application. In order to cosimulate the application and the architecture, a trace-driven simulation approach technique is chosen. Sesame is developed in the context of the Artemis project (architectures and methods for embedded media systems) [13].

The MILAN (model-based integrated simulation) framework is a design space exploration tool that works at different levels of abstraction [14]. Following the Y-chart approach [11], MILAN uses hierarchical dataflow graphs including function alternatives. The architecture template can be defined at different levels of detail. The hierarchical design space exploration starts at the system level and uses rough estimation and symbolic methods based on ordered binary decision diagrams to prune the search space. After reducing the search space, a more fine grained estimation is performed for the remaining designs, reducing the search space even more. At the end, at most ten designs are evaluated by cycle-accurate trace-driven simulation. MILAN needs user interaction to perform decision making during exploration.

In [15], Kianzad and Bhattacharyya propose a framework called CHARMED (cosynthesis of hardware-software multimode embedded systems) for the automatic design space exploration for periodic multimode embedded systems. The input specification is given by several task graphs where each task graph is associated to one of M modes. Moreover, a period for each task graph is given. Associated with the vertices and edges in each task graph, there are attributes like memory requirement and worst case execution time. Two kinds of resources are distinguished, processing elements and communication resources. Kianzad and Bhattacharyya use an approach based on SPEA2 [16] with *constraint dominance*, a similar optimization strategy as implemented by our SystemCoDesigner.

Balarin et al. [17] propose Metropolis, a design space exploration framework which integrates tools for simulation, verification, and synthesis. Metropolis is an infrastructure to help designers to cope with the difficulties in large system designs by allowing the modeling on different levels of detail and supporting refinement. The applications are modeled by a metamodel consisting of sequential processes communicating via the so-called *media*. A medium has variables and functions where the variables are only allowed to be changed by the functions. From the application model a sequence of event vectors is extracted representing a partial execution order. Nondeterminism is allowed in application modeling. The architecture again is modeled by the metamodel, where media are resources and processes representing services (a collection of functions). Deriving the sequence of event vectors results in a nondeterministic execution order of all functions. The mapping is performed by intersecting both event sequences. Scheduling decisions on shared resources are resolved by the so-called *quantity managers* which annotate the events. That way, quantity managers can also be used to associate other properties with events, like power consumption. In contrast to SystemCoDesigner, Metropolis is not concerned with automatic design space exploration. It supports refinement and abstraction, thus allowing top-down and bottom-up methodologies with a meet in the middle approach. As Metropolis is a framework based on a metamodel implementing the Y-chart approach, many system-level design methodologies, including SystemCoDesigner, may be represented in Metropolis.

Finally, some approaches exist to map digital signal processing algorithms automatically to an FPGA platform. *Compaan/Laura* [18] automatically converts a Matlab loop program into a KPN network. This process network can be transformed into a hardware/software system by instantiating IP cores and connecting them with FIFOs. Special software routines take care of the hardware/software communication.

Whereas [18] uses a computer system together with a PCI FPGA board for implementation, [19] automates the generation of a SoC (system on chip). For this purpose, the user has to provide a platform specification enumerating the available microprocessors and communication infrastructure. Furthermore, a mapping has to be provided specifying which process of the KPN graph is executed on which processor unit. This information allows the *ESPAM* tool to assemble a complete system including different communication modules as buses and point-to-point communication. The Xilinx *EDK* tool is used for final bitstream generation.

Whereas both *Compaan/Laura/ESPAM* and SystemCoDesigner want to simplify and accelerate the design of complex hardware/software systems, there are significant differences. First of all, *Compaan/Laura/ESPAM* uses Matlab loop programs as input specification, whereas SystemCoDesigner bases on SystemC allowing for both simulation and automatic hardware generation using behavioral compilers. Furthermore, our specification language SysteMoC is not restricted to KPN, but allows to represent different models of computation.

ESPAM provides a flexible platform using generic communication modules like buses, cross-bars, point-to-point communication, and a generic communication controller. SystemCoDesigner currently restricts to extended FIFO communication allowing out-of-order reads and writes.

Additionally our approach tightly includes automatic design space exploration, estimating the achievable system performance. Starting from an architecture template, a subset of resources is selected in order to obtain an efficient implementation. Such a design point can be automatically translated into a system on chip.

Another very interesting approach based on UML is presented in [20]. It is called Koski and as SystemCoDesigner, it is dedicated to the automatic SoC design. Koski follows the Y-chart approach. The input specification is given as Kahn process networks modeled in UML. The Kahn processes are modeled using Statecharts. The target architecture consists of the application software, the platform-dependent and platform-independent software, and synthesizable communication and processing resources. Moreover, special functions for application distribution are included, that is, interprocess communication for multiprocessor systems. During design space exploration, Koski uses simulation for performance evaluation. Also, Koski has many similarities with SystemCoDesigner, there are major differences. In comparison to SystemCoDesigner, Koski has the following advantages. It supports a network communication which is more platform-independent than the SystemCoDesigner approach. It is also somehow more flexible by supporting a real-time operating System (RTOS) on the CPU. However, there are many advantages when using SystemCoDesigner. (1) SystemCoDesigner permits the specification directly in SystemC and automatically extracts the underlying model of computation. (2) The architecture specification in SystemCoDesigner is not limited to a shared communication medium, it also allows for optimized point-to-point communication. The main advantage of the SystemCoDesigner is its multiobjective design space exploration which allows for optimizing several objectives simultaneously.

The Ptolemy II project [21] was started in 1996 by the University of California, Berkeley. Ptolemy II is a software infrastructure for modeling, analysis, and simulation of embedded systems. The focus of the project is on the integration of different models of computation by the so-called *hierarchical heterogeneity*. Currently, supported MoCs are continuous time, discrete event, synchronous dataflow, FSM, concurrent sequential processes, and process networks. By coupling different MoCs, the designer has the ability to model, analyze, or simulate heterogeneous systems. However, as different actors in Ptolemy II are written in JAVA, it is limited in its usability of the specification for generating efficient hardware/software implementations including hardware and communication synthesis for SoC platforms. Moreover, Ptolemy II does not support automatic design space exploration.

The Signal Processing Worksystem (SPW) from Cadence Design Systems, Inc., is dedicated to the modeling and analysis of signal processing algorithms [22]. The underlying model is based on static and dynamic dataflow models. A hierarchical composition of the actors is supported. The actors themselves can be specified by several different models like SystemC, Matlab, C/C++, Verilog, VHDL, or the design library from SPW. The main focus of the design flow is on simulation and manual refinement. No explicit mapping between application and architecture is supported.

CoCentric System Studio is based on languages like C/C++, SystemC, VHDL, Verilog, and so forth, [23]. It allows for algorithmic and architecture modeling. In System Studio, algorithms might be arbitrarily nested dataflow models and FSMs [24]. But in contrast to Ptolemy II, CoCentric allows hierarchical as well as parallel combinations, what reduces the analysis capability. Analysis is only supported for pure dataflow models (deadlock detection, consistency) and pure FSMs (causality). The architectural model is based on the transaction-level model of SystemC and permits the inclusion of other RTL models as well as algorithmic System Studio models and models from Matlab. No explicit mapping between application and architecture is given. The implementation style is determined by the actual encoding a designer chooses for a module.

Beside the modeling and design space exploration aspects, there are several approaches to efficiently represent MoCs in SystemC. The facilities for implementing MoCs in SystemC have been extended by Herrera et al. [25] who have implemented a custom library of channel types like rendezvous on top of the SystemC discrete event simulation ker-

nel. But no constraints have imposed how these new channel types are used by an actor. Consequently, no information about the communication behavior of an actor can be automatically extracted from the executable specification. Implementing these channels on top of the SystemC discrete event simulation kernel curtails the performance of such an implementation. To overcome these drawbacks, Patel and Shukla [26–28] have extended SystemC itself with different simulation kernels for *communicating sequential processes* (CSP), *continuous time* (CT), *dataflow process networks* (PN) dynamic as well as static (SDF), and *finite state machine* (FSM) MoCs to improve the simulation efficiency of their approach.

3. EXPRESSING DIFFERENT MoCs IN SYSTEMC

In this section, we will introduce our library-based approach to actor-based design called SysteMoC [7] which is used for modeling the behavior and as synthesizable subset of SystemC in our SystemCoDesigner design flow. Instead of a monolithic approach for representing an executable specification as done using many design languages, SysteMoC supports an *actor-oriented* design [29, 30] for many dataflow models of computation (MoCs). These models have been applied successfully in the design of digital signal processing algorithms. In this approach, we consider timing and functionality to be orthogonal. Therefore, our design must be modeled in an untimed dataflow MoC. The timing of the design is derived in the design space exploration phase from mapping of the actors to selected resources. Note that the timing given by that mapping in general affects the execution order of actors. In Section 4, we present a mechanism to evaluate the performance of our application with respect to a candidate architecture.

On the other hand, industrial design flows often rely on executable specifications, which have been encoded in design languages which allow unstructured communication. In order to combine both approaches, we propose the SysteMoC library which permits writing an executable specification in SystemC while separating the *actor functionality* from the *communication behavior*. That way, we are able to identify different MoCs modeled in SysteMoC. This enables us to represent different algorithms ranging from simple static operations modeled by *homogeneous synchronous dataflow* (HSDF) [31] up to complex, data-dependent algorithms as run-length entropy encoding modeled as *Kahn process networks* (KPN) [32]. In this paper, an MPEG-4 decoder [33] will be used to explain our system design methodology which encompasses both algorithm types and can hence only be modeled by *heterogeneous* models of computation.

3.1. Actor-oriented model of an MPEG-4 decoder

In actor-oriented design, *actors* are objects which execute concurrently and can only communicate with each other via *channels* instead of method calls as known in object-oriented design. Actor-oriented designs are often represented by bipartite graphs consisting of channels $c \in C$ and actors $a \in A$, which are connected via point-to-point connections from an

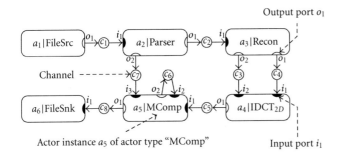

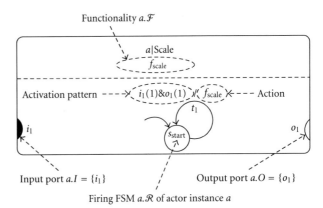

FIGURE 2: The *network graph* of an MPEG-4 decoder. Actors are shown as boxes whereas channels are drawn as circles.

FIGURE 3: Visual representation of the Scale actor as used in the $IDCT_{2D}$ network graph displayed in Figure 4. The Scale actor is composed of *input ports* and *output ports*, its *functionality*, and the *firing FSM* determining the communication behavior of the actor.

actor output port o to a channel and from a channel to an actor input port i. In the following, we call such representations *network graphs*. These network graphs can be extracted directly from the executable SysteMoC specification.

Figure 2 shows the network graph of our MPEG-4 decoder. MPEG-4 [33] is a very complex object-oriented standard for compression of digital videos. It not only encompasses the encoding of the multimedia content, but also the transport over different networks including quality of service aspects as well as user interaction. For the sake of clarity, our decoder implementation restricts to the decompression of a basic video bit-stream which is already locally available. Hence, no transmission issues must be taken into account. Consequently, our bit-stream is read from a file by the FileSrc actor a_1, where $a_1 \in A$ identifies an actor from the set of all actors A.

The Parser actor a_2 analyzes the provided bit-stream and extracts the video data including motion compensation vectors and quantized zig-zag encoded image blocks. The latter ones are forwarded to the reconstruction actor a_3 which establishes the original 8×8 blocks by performing an inverse zig-zag scanning and a dequantization operation. From these data blocks the two-dimensional inverse cosine transform actor a_4 generates the motion-compensated difference blocks. They are processed by the motion compensation actor a_5 in order to obtain the original image frame by taking into account the motion compensation vectors provided by the Parser actor. The resulting image is finally stored to an output file by the FileSnk actor a_6. In the following, we will formally present the SysteMoC modeling concepts in detail.

3.2. SysteMoC concepts

The network graph is the usual representation of an actor-oriented design. It consists of *actors* and *channels*, as seen in Figure 2. More formally, we can derive the following definition.

Definition 1 (network graph). A *network graph* is a directed bipartite graph $g_n = (A, C, P, E)$ containing a set of actors A, a set of channels C, a channel parameter function $P : C \rightarrow \mathbb{N}_\infty \times V^*$ which associates with each channel $c \in C$ its buffer size $n \in \mathbb{N}_\infty = \{1, 2, 3, \ldots, \infty\}$, and also a possibly nonempty sequence $\mathbf{v} \in V^*$ of initial tokens, where

V^* denotes the set of all possible *finite sequences* of tokens $v \in V$ [6]. Additionally, the network graph consists of directed edges $e \in E \subseteq (C \times A.I) \cup (A.O \times C)$ between actor output ports $o \in A.O$ and channels as well as channels and actor input ports $i \in A.I$. These edges are further constraints such that each channel can only represent a point-to-point connection, that is, exactly one edge is connected to each actor port and the in-degree and out-degree of each channel in the graph are exactly one.

Actors are used to model the functionality. An *actor* a is only permitted to communicate with other actors via its actor ports $a.\mathcal{P}$.[1] Other forms of interactor communication are forbidden. In this sense, a network graph is a specialization of the framework concept introduced in [29], which can express an arbitrary connection topology and a set of initial states. Therefore, the corresponding set of framework states Σ is given by the product set of all possible sequences of all channels of the network graph and the single initial state is derived from the channel parameter function P. Furthermore, due to the point-to-point constraint of a network graph, two framework actions λ_1, λ_2 referenced in different framework actors are constrained to only modify parts of the *framework state* corresponding to different network graph channels.

Our actors are composed from *actions* supplying the actor with its data transformation *functionality* and a *firing FSM* encoding, the *communication behavior* of the actor, as illustrated in Figure 3. Accordingly, the state of an actor is also divided into the *functionality state* only modified by the *actions* and the *firing state* only modified by the *firing FSM*. As actions do not depend on or modify the *framework state*

---

[1] We use the "."-operator, for example, $a.\mathcal{P}$, for denoting member access, for example, $\mathcal{P}$, of tuples whose members have been explicitly named in their definition, for example, $a \in A$ from Definition 2. Moreover, this member access operator has a trivial pointwise extension to sets of tuples, for example, $A.\mathcal{P} = \bigcup_{a \in A} a.\mathcal{P}$, which is also used throughout this paper.

their execution corresponds to a sequence of *internal transitions* as defined in [29].

Thus, we can define an actor as follows.

Definition 2 (actor). An *actor* is a tuple $a = (\mathcal{P}, \mathcal{F}, \mathcal{R})$ containing a set of *actor ports* $\mathcal{P} = I \cup O$ partitioned into *actor input ports* I and *actor output ports* O, the *actor functionality* $\mathcal{F}$ and the *firing finite state machine (FSM)* $\mathcal{R}$.

The notion of the *firing FSM* is similar to the concepts introduced in FunState [34] where FSMs locally control the activation of transitions in a Petri Net. In SysteMoC, we have extended FunState by allowing guards to check for available space in output channels before a transition can be executed. The states of the firing FSM are called *firing states*, directed edges between these firing states are called *firing transitions*, or *transitions* for short. The transitions are guarded by *activation patterns* $k = k_{in} \wedge k_{out} \wedge k_{func}$ consisting of (i) predicates k_{in} on the number of available tokens on the input ports called *input patterns*, for example, i(1) denotes a predicate that tests the availability of at least one token on the actor input port i, (ii) predicates k_{out} on the number of free places on the output ports called *output patterns*, for example, o(1) checks if the number of free places of an output is at least one, and (iii) more general predicates k_{func} called *functionality conditions* depending on the *functionality state*, defined below, or the token values on the input ports. Additionally, the transitions are annotated with *actions* defining the *actor functionality* which are executed when the transitions are taken. Therefore, a transition corresponds to a *precise reaction* as defined in [29], where an *input/output pattern* corresponds to an *I/O transition* in the framework model. And an *activation pattern* is always a *responsible trigger*, as actions correspond to a sequence of *internal transitions*, which are independent from the *framework state*.

More formally, we derive the following two definitions.

Definition 3 (firing FSM). The *firing FSM* of an actor $a \in A$ is a tuple $a.\mathcal{R} = (T, Q_{firing}, q_{0firing})$ containing a finite set of *firing transitions* T, a finite set of *firing states* Q_{firing}, and an *initial firing state* $q_{0firing} \in Q_{firing}$.

Definition 4 (transition). A *firing transition* is a tuple $t = (q_{firing}, k, f_{action}, q'_{firing}) \in T$ containing the current firing state $q_{firing} \in Q_{firing}$, an *activation pattern* $k = k_{in} \wedge k_{out} \wedge k_{func}$, the associated *action* $f_{action} \in a.\mathcal{F}$, and the next firing state $q'_{firing} \in Q_{firing}$. The activation pattern k is a Boolean function which determines if transition t can be taken (true) or not (false).

The actor functionality $\mathcal{F}$ is a set of *methods* of an actor partitioned into *actions* used for data transformation and *guards* used in *functionality conditions* of the *activation pattern*, as well as the internal variables of the actor, and their initial values. The values of the internal variables of an actor are called its *functionality state* $q_{func} \in Q_{func}$ and their initial values are called the *initial functionality state* q_{0func}. Actions and guards are partitioned according to two fundamental

differences between them: (i) a guard just returns a Boolean value instead of computing values of tokens for output ports, and (ii) a guard must be side-effect free in the sense that it must not be able to change the functionality state. These concepts can be represented more formally by the following definition.

Definition 5 (functionality). The *actor functionality* of an actor $a \in A$ is a tuple $a.\mathcal{F} = (F, Q_{func}, q_{0func})$ containing a set of *functions* $F = F_{action} \cup F_{guard}$ partitioned into *actions* and *guards*, a set of *functionality states* Q_{func} (possibly infinite), and an *initial functionality state* $q_{0func} \in Q_{func}$.

Example 1. To illustrate these definitions, we give the formal representation of the actor a shown in Figure 3. As can be seen the actor has two *ports*, $\mathcal{P} = \{i_1, o_1\}$, which are partitioned into its set of *input ports*, $I = \{i_1\}$, and its set of *output ports*, $O = \{o_1\}$. Furthermore, the actor contains exactly one *method* $\mathcal{F}.F_{action} = \{f_{scale}\}$, which is the *action* $f_{scale} : V \times Q_{func} \rightarrow V \times Q_{func}$ for generating token $v \in V$ containing scaled IDCT values for the output port o_1 from values received on the input port i_1. Due to the lack of any *internal variables*, as seen in Example 2, the *set of functionality states* $Q_{func} = \{q_{0func}\}$ contains only the *initial functionality state* q_{0func} encoding the scale factor of the actor.

The execution of SysteMoC actors can be divided into three phases. (i) Checking for enabled transitions $t \in T$ in the firing FSM $\mathcal{R}$. (ii) Selecting and executing one enabled transition $t \in T$ which executes the associated actor functionality. (iii) Consuming tokens on the input ports $a.I$ and producing tokens on the output ports $a.O$ as indicated by the associated input and output patterns $t.k_{in}$ and $t.k_{out}$.

3.3. Writing actors in SysteMoC

In the following, we describe the SystemC representation of actors as defined previously. SysteMoC is a C++ class library based on SystemC which provides base classes for actors and network graphs as well as operators for declaring *firing FSMs* for these actors. In SysteMoC, each actor is represented as an instance of an *actor class*, which is derived from the C++ base class smoc_actor, for example, as seen in Example 2, which describes the SysteMoC implementation of the Scale actor already shown in Figure 3. An actor can be subdivided into three parts: (i) actor *input ports* and *output ports*, (ii) actor *functionality*, and (iii) actor *communication behavior* encoded explicitly by the *firing FSM*.

Example 2. SysteMoC code for the Scale actor being part of the MPEG-4 decoder specification.

```
00 class Scale: public smoc_actor {
01 public:
02    // Input  port declaration
03    smoc_port_in<int>  i1;
04    // Output port declaration
05    smoc_port_out<int> o1;
06 private:
```

```
07    // Actor parameters
08    const int G, OS;
09
10    // functionality
11    void scale() { o1[0] = OS
12                  + (G * i1[0]); }
13
14    // Declaration of firing FSM states
15    smoc_firing_state start;
16  public:
17    // The actor constructor is responsible
18    // for declaring the firing FSM and
19    // initializing the actor
20    Scale(sc_module_name name, int G, int OS)
21      : smoc_actor(name, start),
22        G(G), OS(OS) {
23      // start state consists of
24      // a single self loop
25      start =
26        // input pattern requires at least
27        // one token in the FIFO connected
28        // to input port i1
29        (i1.getAvailableTokens() >= 1) >>
30        // output pattern requires at least
31        // space for one token in the FIFO
32        // connected to output port o1
33        (o1.getAvailableSpace()  >= 1) >>
34        // has action Scale::scale and
35        // next state start
36        CALL(Scale::scale)           >>
37        start;
38    }
39  };
```

As known from SystemC, we use port declarations as shown in lines 2–5 to declare the input and output ports $a.\mathcal{P}$ for the actor to communicate with its environment. Note that the usage of sc_fifo_in and sc_fifo_out ports as provided by the SystemC library would not allow the separation of actor functionality and communication behavior as these ports allow the actor functionality to *consume tokens* or *produce tokens*, for example, by calling read or write methods on these ports, respectively. For this reason, the SystemoC library provides its own input and output port declarations smoc_port_in and smoc_port_out. These ports can only be used by the actor functionality to peek token values already available or to produce tokens for the actual communication step. The token production and consumption is thus exclusively controlled by the local *firing FSM a.$\mathcal{R}$* of the actor.

The functions $f \in F$ of the actor functionality $a.\mathcal{F}$ and its functionality state $q_{func} \in Q_{func}$ are represented by the class methods as shown in line 11 and by class member variables (line 8), respectively. The *firing FSM* is constructed in the constructor of the actor class, as seen exemplarily for a single transition in lines 25–37. For each transition $t \in \mathcal{R}.T$, the number of required input tokens, the quantity of produced output tokens, and the called function of the actor functionality are indicated by the help of the methods

getAvailableTokens(), getAvailableSpace(), and CALL(), respectively. Moreover, the source and sink state of the firing FSM are defined by the C++-operators = and >>. For a more detailed description of the *firing FSM* syntax, see [7].

3.4. Application modeling using SystemoC

In the following, we will give an introduction to different MoCs well known in the domain of digital signal processing and their representation in SystemoC by presenting the MPEG-4 application in more detail. As explained earlier in this section, MPEG-4 is a good example of today's complex signal processing applications. They can no longer be modeled at a granularity level sufficiently detailed for design space exploration by restrictive MoCs like synchronous dataflow (SDF) [35]. However, as restrictive MoCs offer better analysis opportunities they should not be discarded for subsystems which do not need more expressiveness. In our SystemoC approach, all actors are described by a uniform modeling language in such a way that for a considered group of actors it can be checked whether they fit into a given restricted MoC. In the following, these principles are shown exemplarily for (i) synchronous dataflow (SDF), (ii) cyclo-static dataflow (CSDF) [36], and (iii) Kahn process networks (KPN) [32].

Synchronous dataflow (SDF) actors produce and consume upon each invocation a static and constant amount of tokens. Hence, their external behavior can be determined statically at compile time. In other words, for a group of SDF actors, it is possible to generate a static schedule at compile time, avoiding the overhead of dynamic scheduling [31, 37, 38]. For homogeneous synchronous dataflow, an even more restricted MoC where each actor consumes and produces exactly one token per invocation and input (output), it is even possible to efficiently compute a rate-optimal buffer allocation [39].

The classification of SystemoC actors is performed by comparing the firing FSM of an actor with different FSM templates, for example, single state with self loop corresponding to the *SDF* domain or circular connected states corresponding to the *CSDF* domain. Due to the SystemoC syntax discussed above, this information can be automatically derived from the C++ actor specification by simply extracting the firing FSM specified in the actor.

More formally, we can derive the following condition: given an actor $a = (\mathcal{P}, \mathcal{F}, \mathcal{R})$, the actor can be classified as belonging to the SDF domain if each transition has the same input pattern and output pattern, that is, for all $t_1, t_2 \in \mathcal{R}.T$: $t_1.k_{in} \equiv t_2.k_{in} \wedge t_1.k_{out} \equiv t_2.k_{out}$.

Our MPEG-4 decoder implementation contains various such actors. Figure 3 represents the firing FSM of a scaler actor which is a simple SDF actor. For each invocation, it reads a frequency coefficient and multiplies it with a constant gain factor in order to adapt its range.

Cyclo-static dataflow (CSDF) actors are an extension of SDF actors because their token consumption and production do not need to be constant but can vary cyclically. For this purpose, their execution is divided into a fixed number

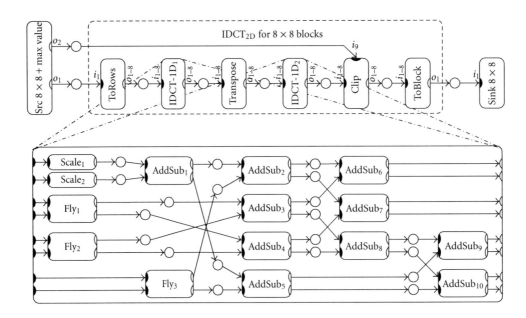

FIGURE 4: The displayed *network graph* is the hierarchical refinement of the $IDCT_{2D}$ actor a_4 from Figure 2. It implements a two-dimensional inverse cosine transformation (IDCT) on 8×8 blocks of pixels. As can be seen in the figure, the two-dimensional inverse cosine transformation is composed of two one-dimensional inverse cosine transformations $IDCT-1D_1$ and $IDCT-1D_2$.

of phases which are repeated periodically. In each phase, a constant number of tokens is written to or read from each actor port. Similar to SDF graphs, a static schedule can be generated at compile time [40]. Although many CSDF graphs can be translated to SDF graphs by accumulating the token consumption and production rates for each actor over all phases, their direct implementation leads mostly to less memory consumption [40].

In our MPEG-4 decoder, the inverse discrete cosine transformation (IDCT), as shown in Figure 4, is a candidate for static scheduling. However, due to the CSDF actor Transpose it cannot be classified as an SDF subsystem. But the contained one-dimensional IDCT is an example of an SDF subsystem, only consisting of actors which satisfy the previously given constraints. An example of such an actor is shown in Figure 3.

An example of a CSDF actor in our MPEG-4 application is the Transpose actor shown in Figure 4 which swaps rows and columns of the 8×8 block of pixels. To expose more parallelism, this actor operates on rows of 8 pixels received in parallel on its 8 input ports i_{1-8}, instead of whole 8×8 blocks, forcing the actor to be a CSDF actor with 8 phases for each of the 8 rows of a 8×8 block. Note that the CSDF actor Transpose is represented in SysteMoC by a firing FSM which contains exactly as many circularly connected firing states as the CSDF actor has execution phases. However, more complex firing FSMs can also exhibit CSDF semantic, for example, due to redundant states in the firing FSM or transitions with the same input and output patterns, the same source and destination firing state but different functionality conditions and actions. Therefore, CSDF actor classification should be performed on a transformed

firing FSM, derived by discarding the *action* and *functionality conditions* from the transitions and performing FSM minimization.

More formally, we can derive the following condition: given an actor $a = (\mathcal{P}, \mathcal{F}, \mathcal{R})$, the actor can be classified as belonging to the CSDF domain if exactly one transition is leaving and entering each firing state, that is, for all $q \in \mathcal{R}.Q_{\text{firing}} : |\{t \in \mathcal{R}.T \mid t.q_{\text{firing}} = q\}| = 1 \wedge |\{t \in \mathcal{R}.T \mid t.q'_{\text{firing}} = q\}| = 1$, and each state of the firing FSM is reachable from the initial state.

Kahn process networks (KPN) can also be modeled in SysteMoC by the use of more general *functionality conditions* in the *activation patterns* of the transitions. This allows to represent data-dependent operations, for example, as needed by the bit-stream parsing as well as the decoding of the variable length codes in the Parser actor. This is exemplarily shown for some transitions of the firing FSM in the Parser actor of the MPEG-4 decoder in order to demonstrate the syntax for using *guards* in the *firing FSM* of an actor. The actions cannot determine presence or absence of tokens, or consume or produce tokens on input or output channels. Therefore, the *blocking reads* of the KPN networks are represented by the blocking behavior of the firing FSM until at least one transition leaving the current firing state is enabled. The behavior of Kahn process networks must be independent from the scheduling strategy. But the scheduling strategy can only influence the behavior of an actor if there is a choice to execute one of the enabled transitions leaving the current state. Therefore, it is possible to determine if an actor a satisfies the KPN requirement by checking for the sufficient condition that all functionality conditions on all transitions leaving a firing state are mutually exclusive, that is, for all $t_1, t_2 \in a.\mathcal{R}.T, t_1.q_{\text{firing}} = t_2.q_{\text{firing}}$: for all $q_{\text{func}} \in a.\mathcal{F}.Q_{\text{func}} : t_1.k_{\text{func}}(q_{\text{func}}) \Rightarrow \neg t_2.k_{\text{func}}(q_{\text{func}}) \wedge t_2.k_{\text{func}}(q_{\text{func}}) \Rightarrow \neg t_1.k_{\text{func}}(q_{\text{func}})$. This guarantees a deterministic behavior of the Kahn process network provided that all actions are also deterministic.

Example 3. Simplified SystemMoC code of the firing FSM analyzing the header of an individual video frame in the MPEG-4 bit-stream.

```
00 class Parser: public smoc_actor {
01 public:
02 // Input port receiving MPEG-4 bit-stream
03 smoc_port_in<int> bits;
04 ...
05 private:
06 // functionality ...
07 // Declaration of guards
08 bool guard_vop_start() const
09  /* code here */
10 bool guard_vop_done () const
11  /* code here */
12 ...
13 // Declaration of firing FSM states
14 smoc_firing_state vol, ..., vop2,
15 vop3, ..., stuck;
16 public:
17 Parser(sc_module_name name)
18 : smoc_actor(name, vol) {
19 ...
20 vop2 = ((bits.getAvailableTokens() >=
21 VOP_START_CODE_LENGTH) &&
22 GUARD(&Parser::guard_vop_done)) >>
23 CALL(Parser::action_vop_done) >>
24 vol
25 | ((bits.getAvailableTokens() >=
26 VOP_START_CODE_LENGTH) &&
27 GUARD(&Parser::guard_vop_start)) >>
28 CALL(Parser::action_vop_start) >>
29 vop3
30 | ((bits.getAvailableTokens() >=
31 VOP_START_CODE_LENGTH) &&
32 !GUARD(&Parser::guard_vop_done) &&
33 !GUARD(&Parser::guard_vop_start)) >>
34 CALL(Parser::action_vop_other) >>
35 stuck;
36 ... // More state declarations
37 }
38 };
```

The data-dependent behavior of the firing FSM is implemented by the guards declared in lines 8–11. These functions can access the values of the input ports without consuming them or performing any other modifications of the functionality state. The GUARD()-method evaluates these guards during determination whether the transition is enabled or not.

4. AUTOMATIC DESIGN SPACE EXPLORATION FOR DIGITAL SIGNAL PROCESSING SYSTEMS

Given an executable signal processing network specification written in SystemMoC, we can perform an automatic design space exploration (DSE). For this purpose, we need additional information, that is, a formal model for the *architecture template* as well as *mapping constraints* for the actors of the SystemMoC application. All these information are captured in a formal model to allow automatic DSE. The task of DSE is to find the best implementations fulfilling the requirements demanded by the formal model. As DSE is often confronted with the simultaneous optimization of many conflicting objectives, there is in general more than a single optimal solution. In fact, the result of the DSE is the so-called *Pareto-optimal set of solutions* [41], or at least an approximation of this set. Beside the task of covering the search space in order to guarantee good solutions, we have to consider the task of evaluating a single design point. In the design of FPGA implementations, the different objectives to minimize are, namely, the number of required look-up tables (LUTs), block RAMs (BRAMs), and flip-flops (FFs). These can be evaluated by analytic methods. However, in order to obtain good performance numbers for other especially important objectives such as latency and throughput, we will propose a simulation-based approach. In the following, we will present the formal model for the exploration, the automatic DSE using multiobjective evolutionary algorithms (MOEAs), as well as the concepts of our simulation-based performance evaluation.

4.1. Design space exploration using MOEAs

For the automatic design space exploration, we provide a formal underpinning. In the following, we will introduce the so-called *specification graph* [42]. This model strictly separates behavior and system structure: the *problem graph* models the behavior of the digital signal processing algorithm. This graph is derived from the *network graph*, as defined in Section 3, by discarding all information inside the actors as described later on. The architecture template is modeled by the so-called *architecture graph*. Finally, the *mapping edges* associate actors of the problem graph with resources in the architecture graph by a "can be implemented by" relation. In the following, we will formalize this model by using the definitions given in [42] in order to define the task of design space exploration formally.

The application is modeled by the so-called *problem graph* $g_p = (V_p, E_p)$. Vertices $v \in V_p$ model actors whereas edges $e \in E_p \subseteq V_p \times V_p$ represent data dependencies between actors. Figure 5 shows a part of the problem graph corresponding to the hierarchical refinement of the IDCT$_{2D}$ actor a_4 from Figure 2. This problem graph is derived from the network graph by a one-to-one correspondence between network graph actors and channels to problem graph vertices while abstracting from

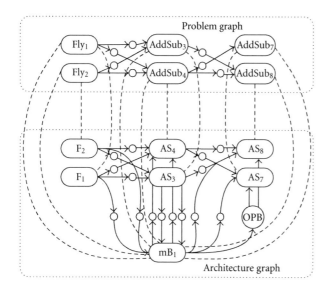

FIGURE 5: Partial specification graph for the IDCT-1D actor as shown in Figure 4. The upper part is a part of the problem graph of the IDCT-1D. The lower part shows the architecture graph consisting of several dedicated resources $\{F_1, F_2, AS_3, AS_4, AS_7, AS_8\}$ as well as a MicroBlaze CPU-core $\{mB_1\}$ and an OPB (open peripheral bus [43]). The dashed lines denote the mapping edges.

actor ports, but keeping the connection topology, that is, $\exists f : g_p.V_p \rightarrow g_n.A \cup g_n.C$, f is a *bijection* : for all $v_1, v_2 \in g_p.V_p$: $(v_1, v_2) \in g_p.E_p \Leftrightarrow (f(v_1) \in g_n.C \Rightarrow \exists p \in f(v_2).I : (f(v_1), p) \in g_n.E) \vee (f(v_2) \in g_n.C \Rightarrow \exists p \in f(v_1).O : (p, f(v_2)) \in g_n.E)$.

The architecture template including functional resources, buses, and memories is also modeled by a directed graph termed *architecture graph* $g_a = (V_a, E_a)$. Vertices $v \in V_a$ model functional resources (RISC processor, coprocessors, or ASIC) and communication resources (shared buses or point-to-point connections). Note that in our approach, we assume that the resources are selected from our component library as shown in Figure 1. These components can be either written by hand in a hardware description language or can be synthesized with the help of high-level synthesis tools such as Mentor CatapultC [8] or Forte Cynthesizer [9]. This is a prerequisite for the later automatic system generation as discussed in Section 5. An edge $e \in E_a$ in the architecture graph g_a models a directed link between two resources. All the resources are viewed as *potentially allocatable* components.

In order to perform an automatic DSE, we need information about the hardware resources that might by allocated. Hence, we annotate these properties to the vertices in the architecture graph g_a. Typical properties are the occupied area by a hardware module or the static power dissipation of a hardware module.

Example 4. For FPGA-based platforms, such as built on Xilinx FPGAs, typical resources are MicroBlaze CPU, open peripheral buses (OPB), fast simplex links (FSLs), or user specified modules representing implementations of actors in the problem graph. In the context of platform-based FPGA

designs, we will consider the number of resources a hardware module is assigned to, that is, for instance, the number of required look-up tables (LUTs), the number of required block RAMs (BRAMs), and the number of required flip-flops (FFs).

Next, it is shown how user-defined mapping constraints representing possible bindings of actors onto resources can be specified in a graph-based model.

Definition 6 (specification graph [42]). A *specification graph* $g_s(V_s, E_s)$ consists of a problem graph $g_p(V_p, E_p)$, an architecture graph $g_a(V_a, E_a)$, and a set of *mapping edges* E_m. In particular, $V_s = V_p \cup V_a$, $E_s = E_p \cup E_a \cup E_m$, where $E_m \subseteq V_p \times V_a$.

Mapping edges relate the vertices of the problem graph to vertices of the architecture graph. The edges represent user-defined mapping constraints in the form of the relation "can be implemented by." Again, we annotate the properties of a particular mapping to an associated mapping edge. Properties of interest are dynamic power dissipation when executing an actor on the associated resource or the worst case execution time (WCET) of the actor when implemented on a CPU-core. In order to be more precise in the evaluation, we will consider the properties associated with the actions of an actor, that is, we annotate for each action the WCET to each mapping edge. Hence, our approach will perform an *actor-accurate binding* using an *action-accurate performance evaluation*, as discussed next.

Example 5. Figure 5 shows an example of a specification graph. The problem graph shown in the upper part is a subgraph of the IDCT-1D problem graph from Figure 4. The architecture graph consists of several dedicated resources connected by FIFO channels as well as a MicroBlaze CPU-core and an on-chip bus called OPB (open peripheral bus [43]). The channels between the MicroBlaze and the dedicated resources are FSLs. The dashed edges between the two graphs are the additional mapping edges E_m that describe the possible mappings. For example, all actors can be executed on the MicroBlaze CPU-core. For the sake of clarity, we omitted the mapping edges for the channels in this example. Moreover, we do not show the costs associated with the vertices in g_a and the mapping edges to maintain clarity of the figure.

In the above way, the model of a specification graph allows a flexible expression of the expert knowledge about useful architectures and mappings. The goal of design space exploration is to find optimal solutions which satisfy the specification given by the specification graph. Such a solution is called a *feasible implementation* of the specified system. Due to the multiobjective nature of this optimization problem, there is in general more than a single optimal solution.

System synthesis

Before discussing automatic design space exploration in detail, we briefly discuss the notion of a *feasible implementation* (cf. [42]). An implementation $\psi = (\alpha, \beta)$, being the result of a system synthesis, consists of two parts: (1) the *allocation* α that indicates which elements of the architecture graph are used in the implementation and (2) the *binding* β, that is, the set of mapping edges which define the binding of vertices in the problem graph to resources of the architecture graph. The task of system synthesis is to determine optimal implementations. To identify the feasible region of the design space, it is necessary to determine the set of *feasible allocations* and *feasible bindings*. A *feasible binding* guarantees that communications demanded by the actors in the problem graph can be established in the allocated architecture. This property makes the resulting optimization problem hard to be solved. A *feasible allocation* is an allocation α that allows at least one feasible binding β.

Example 6. Consider the case that the allocation of vertices in Figure 5 is given as $\alpha = \{mB_1, OPB, AS_3, AS_4\}$. A feasible binding can be given by $\beta = \{(Fly_1, mB_1), (Fly_2, mB_1), (AddSub_3, AS_3), (AddSub_4, AS_4), (AddSub_7, mB_1), (AddSub_8, mB_1)\}$. All channels in the problem graph are mapped onto the OPB.

Given the implementation ψ, some properties of ψ can be calculated. This can be done analytically or simulation-based.

The optimization problem

Beside the problem of determining a single feasible solution, it is also important to identify the set of optimal solutions. This is done during automatic design space exploration (DSE). The task of automatic DSE can be formulated as a *multiobjective combinatorial optimization problem*.

Definition 7 (automatic design space exploration). The task of *automatic design space exploration* is the following multiobjective optimization problem (see, e.g., [44]) where without loss of generality, only minimization problems are assumed here:

$$\text{minimize } f(x),$$
$$\text{subject to :}$$
$$x \text{ represents a feasible implementation } \psi, \quad (1)$$
$$c_i(x) \leq 0, \quad \forall i \in \{1, \dots, q\},$$

where $x = (x_1, x_2, \dots, x_m) \in X$ is the *decision vector*, X is the *decision space*, $f(x) = (f_1(x), f_2(x), \dots, f_n(x)) \in Y$ is the *objective function*, and Y is the *objective space*.

Here, x is an encoding called *decision vector* representing an implementation ψ. Moreover, there are q constraints $c_i(x)$, $i = 1, \dots, q$, imposed on x defining the set of feasible implementations. The *objective function* f is n-dimensional, that is, n objectives are optimized simultaneously. For example, in embedded system design it is required that the monetary cost and the power dissipation of an implementation are minimized simultaneously. Often, objectives in embedded system design are conflicting [45].

Only those *design points* $x \in X$ that represent a feasible implementation ψ and that satisfy all constraints c_i are in the set of feasible solutions, or for short in the *feasible set* called $X_f = \{x \mid \psi(x) \text{ being feasible} \wedge c(x) \leq 0\} \subseteq X$.

A decision vector $x \in X_f$ is said to be nondominated regarding a set $A \subseteq X_f$ if and only if $\nexists a \in A : a \succ x$ with $a \succ x$ if and only if for all $i : f_i(a) \leq f_i(x)$.[2] A decision vector x is said to be Pareto optimal if and only if x is nondominated regarding X_f. The set of all Pareto-optimal solutions is called the *Pareto-optimal set*, or the *Pareto set* for short.

We solve this challenging multiobjective combinatorial optimization problem by using the state-of-the-art MOEAs [46]. For this purpose, we use sophisticated decoding of the individuals as well as integrated symbolic techniques to improve the search speed [2, 42, 47–49]. Beside the task of covering the design space using MOEAs, it is important to evaluate each design point. As many of the considered objectives can be calculated analytically (e.g., FPGA-specific objectives such as total number of LUTs, FFs, BRAMs), we need in general more time-consuming methods to evaluate other objectives. In the following, we will introduce our approach to a simulation-based performance evaluation in order to assess an implementation by means of latency and throughput.

4.2. Simulation-based performance evaluation

Many system-level design approaches rely on application modeling using static dataflow models of computation for signal processing systems. Popular dataflow models are SDF and CSDF or HSDF. Those models of computation allow for static scheduling [31] in order to assess the latency and throughput of a digital signal processing system. On the other hand, the modeling restrictions often prohibit the representation of complex real-world applications, especially if data-dependent control flow or data-dependent actor activation is required. As our approach is not limited to static dataflow models, we are able to model more flexible and complex systems. However, this implies that the performance evaluation in general is not any longer possible through static scheduling approaches.

As synthesizing a hardware prototype for each design point is also too expensive and too time-consuming, a methodology for analyzing the system performance is needed. Generally, there exist two options to assess the performance of a design point: (1) by simulation and (2) by analytical methods. Simulation-based approaches permit a more detailed performance evaluation than formal analyses as the behavior and the timing can interfere as is the case when using nondeterministic merge actors. However, simulation-based approaches reveal only the performance for certain stimuli. In this paper, we focus on a simulation-based performance evaluation and we will show how to generate efficient SystemC simulation models for each design point during DSE automatically.

Our performance evaluation concept is as follows: during design space exploration, we assess the performance of each

---

[2] Without loss of generality, only minimization problems are considered.

feasible implementation with respect to a given set of stimuli. For this purpose, we also model the architecture in SystemC by means of the so-called *virtual processing components* [50]: for each activated vertex in the architecture graph, we create such a virtual processing component. These components are called *virtual* as they are not able to perform any computation but are only used to simulate the delays of actions from actors mapped onto these components. Thus, our simulation approach is called virtual processing components.

In order to simulate the timing of the given SysteMoC application, the actors are mapped onto the virtual processing components according to the binding β. This is established by augmenting the end of all actions $f \in a.\mathcal{F}.F_{\text{action}}$ of each actor $a \in g_{\text{n}}.A$ with the so-called compute function calls. In the simulation, these function calls will block an actor until the corresponding virtual processing components signal the end of the computation. Note that this end time generally depends on (i) the WCET of an action, (ii) other actors bound onto the same virtual processing component, as well as (iii) the stimuli used for simulation. In order to simulate effects of resource contention and resolve resource conflicts, a scheduling strategy is associated with each virtual processing component. The scheduling strategy might be either preemptive or nonpreemptive, like *first come first served, round robin, priority based* [51].

Beside modeling WCETs of each action, we are able to model functional pipelining in our simulation approach. This is established by distinction of WCET and the so-called *data introduction interval* (DII). In this case, resource contention is only considered during the DII. The difference between WCET and DII is an additional delay for the production of output tokens of a computation and does not occupy any resources.

Example 7. Figure 6 shows an example for modeling preemptive scheduling. Two actors, AddSub$_7$ and AddSub$_8$, perform compute function calls on the instantiated MicroBlaze processor mB$_1$. We assume in this example that the MicroBlaze applied a priority-based scheduling strategy for scheduling all actor action execution requests that are bound to the MicroBlaze processor. We also assume that the actor AddSub$_7$ has a higher priority than the actor AddSub$_8$. Thus, the execution of the action f_{addsub} of the AddSub$_7$ actor preempts the execution of the action f'_{addsub} of the AddSub$_8$ actor. Our VPC framework provides the necessary interface between virtual processing components and schedulers: the virtual processing component notifies the scheduler about each compute function call while the scheduler replies with its scheduling decision.

The performance evaluation is performed by a combined simulation, that is, we simulate the functionality and the timing in one single simulation model. As a result of the SystemC-based simulation, we get traces logged during the simulation, showing the activation of actions, the start times, as well as the end times. These traces are used to assess the performance of an implementation by means of *average latency* and *average throughput*. In general, this approach leads

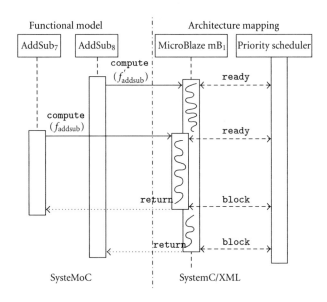

FIGURE 6: Example of modeling preemptive scheduling within the concept of virtual processing components [50]: two actor actions compete for the same virtual processing component by compute function calls. An associated *scheduler* resolves the conflict by selecting the action to be executed.

to very precise simulation results according to the level of abstraction, that is, *action accuracy*.

Compared to other approaches, we support a detailed performance evaluation of heterogeneous multiprocessor architectures supporting arbitrary preemptive and nonpreemptive scheduling strategies, while needing almost no source code modifications. The approach given in [52, 53] allows for modeling of real-time scheduling strategies by introducing a real-time operating system (RTOS) module based on SystemC. Therefore, each atomic operation, for example, any code line, is augmented by an await() function call within all software tasks. Each of those function calls enforces a scheduling decision, also known as *cooperative scheduling*. On top of those predetermined breaking points, the RTOS module emulates a preemptive scheduling policy for software tasks running on the same RTOS module. Another approach found in [54] motivates the so-called *virtual processing units* (VPU) for representing processors. Each VPU supports only a priority-based scheduling strategy. Software processes are modeled as *timed communication extended finite state machines* (tCEFSM). Each state transition of a tCEFSM represents an atomic operation and consumes a fixed amount of processor cycles. The modeling of time is the main limitation of this approach, because each transition of a tCEFSM requires the same number of processor cycles. Our VPC framework overcomes those limitations by combining (i) action-accurate, (ii) resource-accurate, and (iii) contention- and scheduling-accurate timing simulation.

In the Sesame framework [12] a *virtual processor* is used to map an event trace to a SystemC-based transaction level architecture simulation. For this purpose, the application code given as a Kahn process network is annotated with *read*, *write*, and *execute* statements. While executing the Kahn application, traces of *application events* are generated and passed to the *virtual processor*. Computational events (*execute*) are dispatched directly by the *virtual processor* which simulates the timing and communication events (*read*, *write*) are passed to a transaction level SystemC-based architecture simulator. As the scheduling of an event trace in a *virtual processor* does not affect the application, the Sesame framework does not support modeling of time-dependent application behavior. In our VPC framework, application and architecture are simulated in the same simulation-time domain and thus the blocking of a compute function call allows for simulation of time-dependent behavior. Further on, we do not explicitly distinguish between communication and computational execution, instead both types of execution use the compute function call for timing simulation. This abstract modeling of computation and communication delays results in a fast performance evaluation, but does not reveal the details of a transaction level simulation.

One important aspect of our design flow is that we can generate these efficient simulation models automatically. This is due to our SysteMoC library.[3] As we have to control the three phases in the simulation as discussed in Section 3.2, we can introduce the compute function calls directly at the end of phase (ii), that is, no additional modifications of the source code are necessary when using SysteMoC.

In summary, the advantages of virtual processing components are (i) a clear separation between model of computation and model of architecture, (ii) a flexible mapping of the application to the architecture, (iii) a high level of abstraction, and (iv) the combination of functional simulation together with performance simulation.

While performing design space exploration, there is a need for a rapid performance evaluation of different allocations α and bindings β. Thus, the VPC framework was designed for a fast simulation model generation. Figure 7 gives an overview of the implemented concepts. Figure 7(a) shows the implementation $\psi = (\alpha, \beta)$ as a result of the automatic design space exploration. In Figure 7(b), the automatically generated VPC simulation model is shown. The so-called *Director* is responsible for instantiating the virtual processing components according to a given allocation α. Moreover, the binding β is performed by the *Director*, in mapping each SysteMoC actor compute function call to the bound virtual processing components.

Before running the simulation, the *Director* is configured with the necessary information, that is, implementation which should be evaluated. Finally, the *Director* manages the mapping parameters, that is, WCETs and DII of the actions in order to control the simulation times. The configuration is performed through an .xml-file omitting unnecessary recompilations of the simulation model for each design point and, thus, allowing for a fast performance evaluation of large populations of implementations.

---

[3] VPC can also be used together with plain SystemC modules.

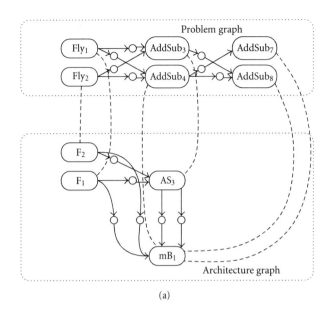

(a)

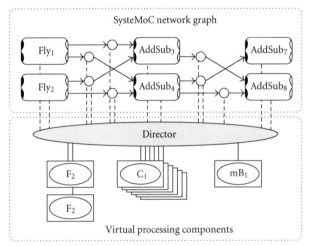

(b)

FIGURE 7: Approach to (i) action-accurate, (ii) resource-accurate, and (iii) contention- and scheduling-accurate simulation-based performance evaluation. (a) An example of one *implementation* as result of the automatic DSE, and (b) the corresponding VPC simulation model. The *Director* constructs the virtual processing components according to the allocation α. Additionally, the *Director* implements the binding of SysteMoC actors onto the virtual processing components according to a specific binding β.

5. AUTOMATIC SYSTEM GENERATION

The result of the automatic design space exploration is a set of nondominated solutions. From these solutions, the designer can select one implementation according to additional requirements or preferences. This process is known as *decision making* in multiobjective optimization.

In this section, we show how to automatically generate a hardware/software implementation for FPGA-based SoC platforms according to the selected allocation and binding. For this purpose, three tasks must be performed: (1) generate the allocated hardware modules, (2) generate the necessary software for each allocated processor core including action code, communication code, and finally scheduling code, and (3) insert the communication resources establishing software/software, hardware/hardware, as well as hardware/software communication. In the following, we will introduce our solution to these three tasks. Moreover, the efficiency of our provided communication resources will be discussed in this section.

5.1. Generating the architecture

Each implementation $\psi = (\alpha, \beta)$ produced by the automatic DSE and selected by the designer is used as input to our automatic system generator for FPGA-based SoC platforms. In our following case study, we specify the system generation flow for Xilinx FPGA platforms only. Figure 8 shows the general flow for Xilinx platforms. The architecture generation is threefold: first, the system generator automatically generates the MicroBlaze subsystems, that is, for each allocated CPU resource, a MicroBlaze subsystem is instantiated. Second, the system generator automatically inserts the allocated IP cores. Finally, the system generator automatically inserts the communication resources. The result of this architecture generation is a hardware description file (.mhs-file) in case of the Xilinx EDK (embedded development Kit [55]) toolchain. In the following, we discuss some details of the architecture generation process.

According to these three above-mentioned steps, the resources in the architecture graph can be classified to be of type *MicroBlaze*, *IP core*, or *Channel*. In order to allow a hardware synthesis of this architecture, the vertices in the architecture graph contain additional information, as, for example, the memory sizes of the MicroBlazes or the names and versions of VHDL descriptions representing the IP cores.

Beside the information stored in the architecture graph, information of the SysteMoC application must be considered during the architecture generation as well. A vertex in the problem graph is either of type *Actor* or of type *Fifo*. Consider the Fly$_1$ actor and communication vertices between actors shown in Figure 8, respectively. A vertex of type *Actor* contains information about the order and values of constructor parameters belonging to the corresponding SysteMoC actor. A vertex of type *Fifo* contains information about the depth and the data type of the communication channel used in the SysteMoC application. If a SysteMoC actor is bound onto a dedicated IP core, the VHDL/Verilog source files of the IP core must be stored in the component library (see Figure 8). For each vertex of type *Actor*, the mapping of SysteMoC constructor parameters to corresponding VHDL/Verilog generics is stored in an *actor information file* to avoid, for example, name conflicts. Moreover, the mapping of SysteMoC ports to VHDL ports has to be taken into account as they do not have to be necessarily the same.

As the system generator traverses the architecture graph, it starts for each vertex in the architecture graph of type *MicroBlaze* or *IP core* the corresponding subsynthesizer which produces an entry in the EDK architecture file. The vertices which are mapped onto a MicroBlaze are determined and registered for the automatic software generation, as discussed in the next section.

After instantiating the MicroBlaze cores and the IP cores, the final step is to insert the communication resources. These communication resources are taken from our platform-specific communication library (see Figure 8). We will discuss this communication library in more detail in Section 5.3. For now, we only give a brief introduction. The software/software communication of SysteMoC actors is done through special SysteMoC software FIFOs by exchanging data within the MicroBlaze by reads and writes to local memory buffers. For hardware/hardware communication, that is, communication between IP cores, we insert the special so-called SysteMoC FIFO which allows, for example, nondestructive reads. It will be discussed in more detail in Section 5.3. The hardware/software communication is mapped on special SysteMoC hardware/software FIFOs. These FIFOs are connected to instantiated fast simplex link (FSL) ports of a MicroBlaze core. Thus, outgoing and incoming communication of actors running on a MicroBlaze use the corresponding implementation which transfers data via the FSL ports of MicroBlaze cores. In case of transmitting data from an IP core to a MicroBlaze, the so-called *smoc2fsl-bridge* transfers data from the IP core to the corresponding FSL port. The opposite communication direction instantiates an *fsl2smoc-bridge*.

After generating the architecture and running our software synthesis tool for SysteMoC actors mapped onto each MicroBlaze, as discussed next, several Xilinx implementation tools are started which produce the platform specific bit file by using several Xilinx synthesis tools including *mb-gcc*, *map*, *par*, *bitgen*, *data2mem*, and so on. Finally, the bit file can be loaded on the FPGA platform and the application can be run.

5.2. Generating the software

In case multiple actors are mapped onto one CPU-core, we generate the so-called *self-schedules*, that is, each actor is tested round robin if it has a fireable action. For this purpose, each SysteMoC actor is translated into a C++ class. The actor functionality $\mathcal{F}$ is copied to the new C++ class, that is, member variables and functions. Actor ports $\mathcal{P}$ are replaced by pointers to the SysteMoC software FIFOs. Finally, for the firing FSM $\mathcal{R}$, a special method called `fire` is generated. Thus, the `fire` method checks the activation of the actor and performs if possible an activated state transition.

To finalize the software generation, instances of each actors corresponding C++ class as well as instances of required SysteMoC software FIFOs are created in a top-level file. In our default implementation, the main function of each CPU-core consists of a `while(true)` loop which tries to execute each actor in a round robin discipline (self-scheduling).

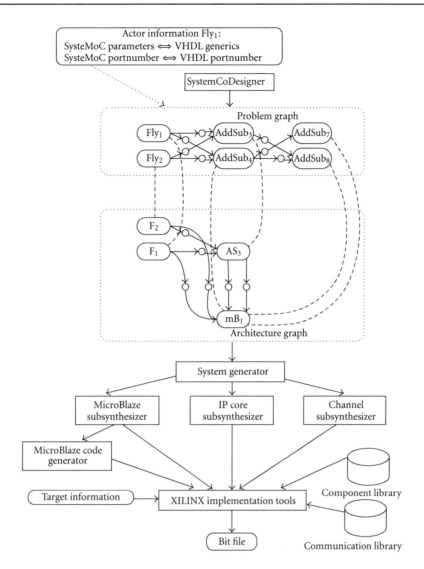

FIGURE 8: Automatic system generation: starting with the selected implementation within the automatic DSE, the system generator automatically generates the MicroBlaze subsystems, inserts the allocated IP cores, and finally connects these functional resources by communication resources. The bit file for configuring the FPGA is automatically generated by an additional software synthesis step and by using the Xilinx design tools, that is, the embedded development kit (EDK) [55] toolchain.

The proposed software generation shows similarities to the software generations discussed in [56, 57]. However, in future work our approach has the potential to replace the above self-scheduling strategy by more sophisticated dynamic scheduling strategies or even optimized static or quasi-static schedules by analyzing the firing FSMs.

In future work we can additionally modify the software generation for DSPs to replace the actor functionality $\mathcal{F}$ with an optimized function provided by DSP vendors, similar as described in [58].

5.3. SysteMoC communication resources

In this section, we introduce our communication library which is used during system generation. The library supports software/software, hardware/hardware, as well as hardware/software communication. All these different kinds of communication provide the same interface as shown in Table 1. This is a quite intuitive interface definition that is similar to interfaces used in other works, like, for example, [59]. In the following, we call each communication resource which implements our interface a SysteMoC FIFO.

The SysteMoC FIFO communication resource provides three different services. They store data, transport data, and synchronize the actors via availability of tokens, respectively, buffer space. The implementation of this communication resource is not limited to be a simple FIFO, it may, for example, consist of two hardware modules that communicate over a bus. In this case, one of the modules would implement the read interface, the other one the write interface.

To be able to store data in the SysteMoC FIFO, it has to contain a buffer. Depending on the implementation, this

TABLE 1: SysteMoC FIFO interface.

Operation	Behavior
rd_tokens()	Returns how many tokens can be read from the SysteMoC-FIFO (available tokens).
wr_tokens()	Returns how many tokens can be written into the SysteMoC-FIFO (free tokens).
read(offset)	Reads a token from a given offset relative to the first available token. The read token is not removed from the SysteMoC-FIFO.
write(offset, value)	Writes a token to a given offset relative to the first free token. The written token is not made available.
rd_commit(count)	Removes count tokens from the SysteMoC-FIFO.
wr_commit(count)	Makes count tokens available for reading.

buffer may also be distributed over different modules. Of course, it would be possible to optimize the buffer sizes for a given application. However, this is currently not supported in SystemCoDesigner. The network graph given by the user contains buffer sizes.

As can be seen from Table 1, a SysteMoC FIFO is more complex than a simple FIFO. This is due to the fact that simple FIFOs do not support nonconsuming read operations for guard functions and that SysteMoC FIFOs must be able to commit more than one read or written token.

For actors that are implemented in software, our communication library supports an efficient software implementation of the described interface. These SysteMoC software FIFOs are based on shared memory and thus allow actors to use a low-overhead communication. For hardware/hardware communication, there is an implementation for Xilinx FPGAs in our communication library. This SysteMoC hardware FIFO uses embedded Block RAM (BRAM) and allows to write and read tokens concurrently every clock cycle. Due to the more complex operations of the SysteMoC hardware FIFO, they are larger than simple native FIFOs created with, for example, CORE Generator for Xilinx FPGAs.

For a comparison, we synthesized different 32 bit wide SysteMoC hardware FIFOs as well as FIFOs generated by Xilinx's CORE generator for an Xilinx XC2VP30 FPGA. The CORE generator FIFOs are created using the synchronous FIFO v5.0 generator without any optional ports and using BRAM. Figure 9 shows the number of occupied flip-flops (FFs) and 4-input look-up tables (LUTs) for FIFOs of different depths. The number of used Block RAMs only depends on the depth and the width of the FIFOs and thus does not vary between SysteMoC and CORE Generator FIFOs.

As Figure 9 shows, the maximum overhead for 4096 tokens depth FIFOs is just 12 FFs and 33 4-input LUTs. Compared to the required 8 BRAMs, this is a very small overhead. Even the maximum clock rates for these FIFOs are very similar and with more than 200 MHz about 4 times higher than typically required.

The last kind of communication resources is the SysteMoC hardware/software FIFOs. Our communication library

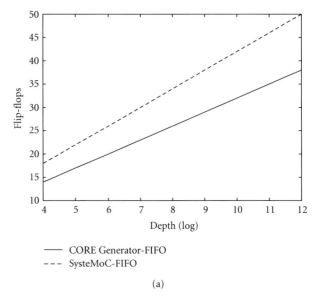

(a)

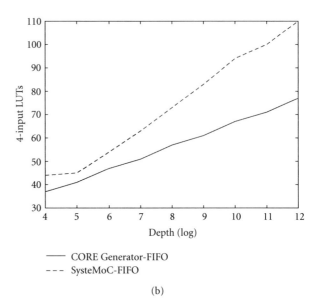

(b)

FIGURE 9: Comparison of (a) flip-flops (FF) and (b) 4-input look-up tables (LUTs) for SysteMoC hardware FIFOs and simple native FIFOs generated by Xilinx's CORE Generator.

supports two different types called smoc2fsl-bridge and fsl2smoc-bridge. As the name suggests, the communication is done via fast simplex links (FSLs). In order to provide the SysteMoC FIFO interface as shown in Table 1 to the software, there is a software driver with some local memory to implement this interface and access the FSL ports of the MicroBlazes. The smoc2fsl-bridge and fsl2smoc-bridge are required adapters to connect hardware SysteMoC FIFOs to FSL ports. Therefore, the smoc2fsl-bridge reads values from a connected SysteMoC FIFO and writes them to the FSL port. On the other side, the fsl2smoc-bridge allows to transfer data from a FSL port to a hardware SysteMoC FIFO.

6. RESULTS

In this section, we demonstrate first results of our design flow by applying it to the two-dimensional inverse discrete cosine transform (IDCT_{2D}) being part of the MPEG-4 decoder model. Principally, this encompasses the following tasks. (i) Estimation of the attributes, like number of flip-flops or execution delays required for the automatic design space exploration (DSE). (ii) Generation of the specification graph and performing the automatic DSE. (iii) Selection of design points due to the designer's preferences and their automatic translation into a hardware/software system with the methods described in Section 5. In the following, we will present these issues as implemented in SystemCoDesigner in more detail using the IDCT_{2D} example. Moreover, we will analyze the accuracy between design parameters estimated by our simulation model and the implementation as a first step towards an optimized design flow. By restricting to the IDCT_{2D} with its data independent behavior, comparison between the VPC estimates and the measured values of the real implementations can be performed particularly well. This allows to clearly show the benefits of our approach as well as to analyze the reasons for observed differences.

6.1. Determination of the actor attributes

As described in Section 4, automatic design space exploration (DSE) selects implementation alternatives based on different objectives as, for example, the number of hardware resources or achieved throughput and latency. These objectives are calculated based on the information available for a single actor action or hardware module. For the hardware modules, we have taken into account the number of flip-flops (FFs), look-up tables (LUTs), and block RAM (BRAM). As our design methodology allows for parameterized hardware IP cores, and as the concrete parameter values influence the required hardware resources, the latter ones are determined by generating an implementation where each actor is mapped to the corresponding hardware IP core. A synthesis run with a tool like Xilinx XST then delivers the required values.

Furthermore, we have derived the execution time for each actor action if implemented as hardware module. Whereas the hardware resource attributes differ with the actor parameters, the execution times stay constant for our application and can hence be predetermined once for each IP core by VHDL code analysis. Additionally, the software execution time is determined for each action of each SysteMoC actor through processing it by our software synthesis tool (see Section 5.2) and execution on the MicroBlaze processor, stimulated by a test pattern. The corresponding execution times can then be measured using an *instruction set simulator*, a hardware profiler, or a simulation with, for example, *Modelsim* [8].

6.2. Performing automatic design space exploration

To start the design space exploration we need to construct a specification graph for our IDCT_{2D} example which consists of

TABLE 2: Results of a design space exploration running for 14 hours and 18 minutes using a Linux workstation with a 1800 MHz AMD Athlon XP Processor and 1 GB of RAM.

Parameter	Value
Population archive	500
Parents	75
Children	75
Generations	300
Individuals overall	23 000
Nondominated individuals	1 002
Exploration time	14 h 18 min
Overall simulation time	3 h 18 min
Simulation time	0.52 s/individual

about 45 actors and about 90 FIFOs. Starting from the problem graph, an architecture template is constructed, such that a hardware-only solution is possible. In other words, each actor can be mapped to a corresponding dedicated hardware module. For the FIFOs, we allow two implementation alternatives, namely, block RAM (BRAM) based and look-up table (LUT) based. Hence, we force the automatic design space exploration to find the best implementation for each FIFO. Intuitively, large FIFOs should make use of BRAMs as otherwise too many LUTs are required. Small FIFOs on the other hand can be synthesized using LUTs, as the number of BRAMs available in an FPGA is restricted.

To this hardware-only architecture graph, a variable number of MicroBlaze processors are added, so that each actor can also be executed in software. In this paper, we have used a fixed configuration for the MicroBlaze softcore processor including 128 kB of BRAM for the software. Finally, the mapping of the problem graph to this architecture graph is determined in order to obtain the specification graph. The latter one is annotated with the objective attributes determined as described above and serves as input to the automatic DSE.

In our experiments, we explore a five-dimensional design space where throughput is maximized, while latency, number of look-up tables (LUTs), number of flip-flops (FFs), as well as the sum of BRAM and multiplier resources are minimized simultaneously. The BRAM and the multiplier resources are combined to one objective, as they cannot be allocated independently in Xilinx Virtex-II Pro devices. In general, a pair of one multiplier and one BRAM conflict each other by using the same communication resources in a Xilinx Virtex-II Pro device. For some special cases a combined usage of the BRAM-multiplier pair is possible. This could be taken into account by our design space exploration through inclusion of BRAM access width. However, for reasons of clarity this is not considered furthermore in this paper.

Table 2 gives the results of a single run of the design space exploration of the IDCT_{2D}. The exploration has been stopped after 300 generations which corresponds to 14 hours, and 18

TABLE 3: Comparison of the results obtained by estimation during exploration and after system synthesis. The last table line shows the values obtained for an optimized two-dimensional IDCT module generated by the Xilinx CORE Generator, working on 8×8 blocks.

SW-actors	LUT	FF	BRAM/MUL	Throughput (Blocks/s)	Latency (μs/Block)	
	12 436	7 875	85	155 763.23	22.71	Estimation
0	11 464	7 774	85	155 763.23	22.27	Synthesis
	8.5%	1.3%	0%	0%	2.0%	rel. error
	8 633	4 377	85	75.02	65 505.50	Estimation
24	7 971	4 220	85	70.84	71 058.87	Synthesis
	8.3%	3.7%	0%	5.9%	7.8%	rel. error
	3 498	2 345	70	45.62	143 849.00	Estimation
40	3 152	2 175	70	24.26	265 427.68	Synthesis
	11.0%	7.8%	0%	88.1%	45.8%	rel. error
	2 166	1 281	67	41.71	157 931.00	Estimation
44	1 791	1 122	67	22.84	281 616.43	Synthesis
	23.0%	14.2%	0%	82.6%	43.9%	rel. error
	1 949	1 083	67	41.27	159 547.00	Estimation
All	1 603	899	67	22.70	283 619.82	Synthesis
	21.6%	20.5%	0%	81.8%	43.7%	rel. error
0	2 651	3 333	1	781 250.00	1.86	CORE Generator

minutes.[4] This exploration run was made on a typical Linux workstation with a single 1800 MHz AMD Athlon XP Processor and a memory size of 1 GB. Main part of the time was used for simulation and subsequent throughput and latency calculation for each design point using SysteMoC and the VPC framework. More precisely, the accumulated wall-clock time for all individuals is about 3 hours and the accumulated time needed to calculate the performance numbers is about 6 hours, leading to average wall-clock time of 0.52 seconds and 0.95 seconds, respectively. The set of stimuli used in simulation consists of 10 blocks with size of 8×8 pixels. In summary, the exploration produced 23 000 design points over 300 populations, having 500 individuals and 75 children in each population.[5] At the end of the design space exploration, we counted 1,002 non-dominated individuals. Two salient Pareto-optimal solutions are the hardware-only solution and the software-only solution. The hardware-only implementation obtains the best performance with a latency of 22.71 μs/Block and a throughput of one block each 6.42 μs, more than 155.76 Blocks/ms. The software-only solution needs the minimum number of 67 BRAMs and multipliers, the minimum number of 1 083 flip-flops, and the minimum number of 1 949 look-up tables.

6.3. Automatic system generation

To demonstrate our system design methodology, we have selected 5 design points generated by the design space exploration, which are automatically implemented by our system generator tool.

Table 3 shows both the values determined by the exploration tool (estimation), as well as those measured for the implementation (synthesis). Considering the hardware resources, the estimations obtained during exploration are quite close to the results obtained for the synthesized FPGA circuit. The variations can be explained by post synthesis optimizations as, for example, by register duplication or removal, by trimming of unused logic paths, and so forth, which cannot be taken into account by our exploration tool. Furthermore, the size of the MicroBlaze varies with its configuration, as, for example, the number of FSL links. As we have assumed the worst case of 16 used FSL ports per MicroBlaze, this effect can be particularly well seen for the software-only solution, where the influence of the missing FSL links is clearly visible.

Concerning throughput and latency, we have to distinguish two cases: pure hardware implementations and designs including a processor softcore. In the first case, there is a quite good match between the expected values obtained by simulation and the measured ones for the concrete hardware implementation. Consequently, our approach for characterizing each hardware module individually as an input for our actor-based VPC simulation shows to be worthwhile. The observed differences between the measured values and the estimations performed by the VPC framework can be explained by the early communication behavior of several IP cores as explained in Section 6.3.1.

For solutions including software, the differences are more pronounced. This is due to the fact that our simulation is only an approximation of the implementation. In particular, we have identified the following sources for the observed differences: (i) communication processes encompassing more than one hardware resource, (ii) the scheduling overhead caused by software execution, (iii) the execution order caused by different scheduling policies, and (iv) variable

---

[4] Each generation corresponds to a population of several individuals where each individual represents a hardware/software solution of the IDCT$_{2D}$ example.

[5] The initial population started with 500 random generated individuals.

TABLE 4: Overall overhead for the implementations shown in Table 3. The overhead due to scheduling decisions is given explicitly.

SW-actors	Overhead		Throughput (Blocks/s)		Latency (μs/Block)	
	Overall	Sched.	Cor. simulation	Cor. error	Cor. simulation	Cor. error
24	6.9%	0.9%	69.84	1.4%	70 360.36	1.0%
40	43.7%	39.9%	25.68	5.9%	255 504.44	3.7%
44	41.3%	40.2%	24.48	7.2%	269 047.70	4.5%
All	41.0%	41.0%	24.36	7.3%	270 267.58	4.7%

guard and action execution times caused by conditional code statements.

In the following sections, we will shortly review each of the above-mentioned points explaining the discrepancy between the VPC values for throughput and latency and the results of our measurements.

Finally, Section 6.3.5 is dedicated to the comparison of the optimized CORE Generator module and the implementations obtained by our automatic approach.

6.3.1. Early communication of hardware IP cores

The differences occurring for the latency values of the hardware-only solution can be mainly explained by the communication behavior of the IP cores. According to SysteMoC semantics, communication takes only place after having executed the corresponding action. In other words, the consumed tokens are only removed from the input FIFOs after the actor action has been terminated. The equivalent holds for the produced tokens.

For hardware modules, this behavior is not very common. Especially the input tokens are removed from the input FIFOs rather than at the beginning of the action. Hence, this can lead to earlier firing times of the corresponding source actor in hardware than supposed by the VPC simulation. Furthermore, some of the IP cores pass the generated values to the output FIFOs' some clock cycles before the end of the actor action. Examples are, for instance, the actors block2row and transpose. Consequently, the corresponding sink actor can also fire earlier. In the general case, this behavior can lead to variations in both throughput and latency between the estimation performed by the VPC framework and the measured value.

6.3.2. Multiresource communication

For the hardware/software systems, parts of the differences observed between the VPC simulation and the real implementation can be attributed to the communication processes between IP cores and the MicroBlaze. As our SysteMoC FIFOs allow for access to values addressed by an offset (see Section 5.3), it is not possible to directly use the FSL interface provided by the MicroBlaze processor. Instead, a software layer has to be added. Hence, a communication between both a MicroBlaze and an IP core activates the hardware itself as well as the MicroBlaze. In order to represent this behavior correctly in our VPC framework, a communication process between a hardware and a software actor must be mapped

to several resources (multihop communication). As the current version of our SystemCoDesigner does not provide this feature, the hardware/software communication can only be mapped to the hardware FIFO. Consequently, the time which the MicroBlaze spends for the communication is not correctly taken into account and the estimations for throughput and latency performed by the VPC framework are too optimistic.

6.3.3. Scheduling overhead

A second major reason for the discrepancy between the VPC estimations and the real implementations is situated in the scheduling overhead. The latter one is the time required for determination of the next actor which can be executed. Whereas in our simulation performed during automatic design space exploration, this decision can be performed in zero time (simulated time), this is not true any more for implementations running on a MicroBlaze processor. This is because the test whether an actor can be fired requires the verification of all conditions for the next possible transitions of the firing state machine. This results in one or more function calls.

In order to assess the overhead which is not taken into account by our VPC simulation, we evaluated it for each example implementation given in Table 3 by hand. For the software-only solutions, this overhead exactly corresponds to the scheduling decisions, whereas for the hardware/software realizations it encompasses both schedule decisions and communication overhead on the MicroBlaze processor (Section 6.3.2).

The corresponding results are shown in Table 4. It clearly shows that most of the differences between the VPC simulation and measured results are caused by the neglected overhead. However, inclusion of this time overhead is unfortunately not easy to perform, because the scheduling algorithms used for simulation and for the MicroBlaze implementation differ at least in the order by which the activation patterns of the actors are evaluated. Furthermore, due to the abbreviated conditional execution realized in modern compilers, the verification of the transition predicate can take variable time. Consequently, the exact value of the overhead depends on the concrete implementation and cannot be calculated by some means as clearly shown by Table 4.

For our IDCT$_{2D}$ example, this overhead is particularly pronounced, because the model has a very fine granularity. Hence, the neglected times for scheduling and communication do not differ substantially from the action execution times. A possible solution to this problem is to determine a quasi-static schedule [60], whereas many decisions as possible are done during compile time. Consequently, the scheduling overhead would decrease. Furthermore, this would improve the implementation efficiency. Also, in a system-level implementation of the $IDCT_{2D}$ as part of the MPEG-4 decoder, one could draw the conclusion from the scheduling overhead that the level of granularity for actors that are explored and mapped should be increased.

6.3.4. Execution order

As shown in Table 4, most of the differences occurring between estimated and measured values are caused by the scheduling and communication overhead. The staying difference, typically less than 10%, is due to the different actor execution order, because it influences both initialization and termination of the system.

Taking, for instance, the software-only implementation, then at the beginning all FIFOs are empty. Consequently, the workload of the processor is relatively small. Hence, the first 8×8 block can be processed with a high priority, leading to a small latency. As however the scheduler will start to process a new block before the previous one is finished, the system load in terms of number of simultaneously active blocks will increase until the FIFOs are saturated. In other words, different blocks have to share the CPU, hence latency will increase. On the other hand, when the source stops to process blocks, the system workload gets smaller, leading to smaller latency.

These variations in latency depend on the time, when the scheduler starts to process the next block. Consequently, as our VPC simulation and the implementation use different actor invocation order, also the measured performance value can differ. This can be avoided by using a simulation where the CPU only processes one block per time. Hence, the latency of one block is not affected by the arrival of further blocks.

A similar observation can be made for throughput. The latter one meets its final value only after the system is completely saturated, because it is influenced by the increasing and decreasing block latencies caused at the system startup and termination phase, respectively.

By taking this effects into account, we have been able to further reduce the differences between the VPC estimations and the measured values to 1%-2%.

6.3.5. Comparison with optimized core generator module

Efficient implementation of the inverse discrete cosine transform is very challenging and extensively treated in literature (i.e., [61–64]). In order to compare our automatically built implementations with such optimized realizations, Table 3 includes a Xilinx CORE Generator Module performing a two-dimensional cosine transform. It is optimized to Xilinx FPGAs and is hence a good reference for comparison.

Due to the various possible optimizations for efficient implementations of an $IDCT_{2D}$, it can be expected that au-

tomatically generated solutions have difficulties to reach the same efficiency. This is clearly confirmed by Table 3. Even the hardware-only solution is far slower than the Xilinx CORE Generator module.

This can be explained by several reasons. First of all, our current IP library is not already optimized for area and speed, as the major intention of this paper lies in the illustration of our overall system design flow instead of coping with details of IDCT implementation. As a consequence, the IP cores are not pipelined and their communication handshaking is realized in a safe, but slow way. Furthermore, for the sake of simplicity we have abstained from extensive logic optimization in order to reduce chip area.

As a second major reason, we have identified the scheduling overhead. Due to the self-timed communication of the different modules on a very low level (i.e., a *clip* actor just performs a simple minimum determination), a very large overhead occurs due to required FIFOs and communication state machines, reducing system throughput and increasing chip area. This is particularly true, when a MicroBlaze is instantiated slowing down the whole chain. Due to the simple actions, the communication and schedule overhead play an important role. In order to solve this problem, we currently investigate on quasi-static scheduling and actor clustering for more efficient data transport. This, however, is not in the scope of this paper.

7. CONCLUSIONS

In this paper, we have presented a first prototype of SystemCoDesigner, which implements a seamless automatic design flow for digital signal processing systems to FPGA-based SoC platforms. The key advantage of our proposed hardware/software codesign approach is the combination of executable specifications written in SystemC with formal methods. For this purpose, SysteMoC, a SystemC library for actor-based design, is proposed which allows the identification of the underlying model of computation. The proposed design flow includes application modeling in SysteMoC, automatic design space exploration (DSE) using simulation-based performance evaluation, as well as automatic system generation for FPGA-based platforms. We have shown the applicability of our proposed design flow by presenting first results from applying SystemCoDesigner to the design of a two-dimensional inverse discrete cosine transformation ($IDCT_{2D}$). The results have shown that (i) we are able to automatically optimize and correctly synthesize digital signal processing applications written in SystemC and (ii) our performance evaluation during DSE produces good estimations for the hardware synthesis and less-accurate estimations for the software synthesis.

In future work we will add support for different FPGA platforms and extend our component and communication libraries. Especially, we will focus on the support for non-FIFO communication using on-chip buses. Moreover, we will strengthen our design flow by incorporating formal analysis methods, automatic code transformations, as well as verification support.

REFERENCES

[1] M. Gries, "Methods for evaluating and covering the design space during early design development," *Integration, the VLSI Journal*, vol. 38, no. 2, pp. 131–183, 2004.

[2] C. Haubelt, *Automatic model-based design space exploration for embedded systems—a system level approach*, Ph.D. thesis, University of Erlangen-Nuremberg, Erlangen, Germany, July 2005.

[3] OSCI, "Functional Specification for SystemC 2.0," Open SystemC Initiative, 2002, http://www.systemc.org/.

[4] T. Grötker, S. Liao, G. Martin, and S. Swan, *System Design with SystemC*, Kluwer Academic, Norwell, Mass, USA, 2002.

[5] IEEE, *IEEE Standard SystemC Language Reference Manual (IEEE Std 1666-2005)*, March 2006.

[6] E. A. Lee and A. Sangiovanni-Vincentelli, "A framework for comparing models of computation," *IEEE Transactions on Computer-Aided Design of Integrated Circuits and Systems*, vol. 17, no. 12, pp. 1217–1229, 1998.

[7] J. Falk, C. Haubelt, and J. Teich, "Efficient representation and simulation of model-based designs in SystemC," in *Proceedings of the International Forum on Specification & Design Languages (FDL '06)*, pp. 129–134, Darmstadt, Germany, September 2006.

[8] http://www.mentor.com/.

[9] http://www.forteds.com/.

[10] B. Kienhuis, E. Deprettere, K. Vissers, and P. van der Wolf, "An approach for quantitative analysis of application-specific dataflow architectures," in *Proceedings of the IEEE International Conference on Application-Specific Systems, Architectures and Processors (ASAP '97)*, pp. 338–349, Zurich, Switzerland, July 1997.

[11] A. C. J. Kienhuis, *Design space exploration of stream-based dataflow architectures—methods and tools*, Ph.D. thesis, Delft University of Technology, Delft, The Netherlands, January 1999.

[12] A. D. Pimentel, C. Erbas, and S. Polstra, "A systematic approach to exploring embedded system architectures at multiple abstraction levels," *IEEE Transactions on Computers*, vol. 55, no. 2, pp. 99–112, 2006.

[13] A. D. Pimentel, L. O. Hertzberger, P. Lieverse, P. van der Wolf, and E. F. Deprettere, "Exploring embedded-systems architectures with artemis," *Computer*, vol. 34, no. 11, pp. 57–63, 2001.

[14] S. Mohanty, V. K. Prasanna, S. Neema, and J. Davis, "Rapid design space exploration of heterogeneous embedded systems using symbolic search and multi-granular simulation," in *Proceedings of the Joint Conference on Languages, Compilers and Tools for Embedded Systems: Software and Compilers for Embedded Systems*, pp. 18–27, Berlin, Germany, June 2002.

[15] V. Kianzad and S. S. Bhattacharyya, "CHARMED: a multi-objective co-synthesis framework for multi-mode embedded systems," in *Proceedings of the 15th IEEE International Conference on Application-Specific Systems, Architectures and Processors (ASAP '04)*, pp. 28–40, Galveston, Tex, USA, September 2004.

[16] E. Zitzler, M. Laumanns, and L. Thiele, "SPEA2: improving the strength pareto evolutionary algorithm for multiobjective optimization," in *Evolutionary Methods for Design, Optimization and Control*, pp. 19–26, Barcelona, Spain, 2002.

[17] F. Balarin, Y. Watanabe, H. Hsieh, L. Lavagno, C. Passerone, and A. Sangiovanni-Vincentelli, "Metropolis: an integrated electronic system design environment," *Computer*, vol. 36, no. 4, pp. 45–52, 2003.

[18] T. Stefanov, C. Zissulescu, A. Turjan, B. Kienhuis, and E. Deprettere, "System design using Khan process networks: the Compaan/Laura approach," in *Proceedings of Design, Automation and Test in Europe (DATE '04)*, vol. 1, pp. 340–345, Paris, France, February 2004.

[19] H. Nikolov, T. Stefanov, and E. Deprettere, "Multi-processor system design with ESPAM," in *Proceedings of the 4th International Conference on Hardware/Software Codesign and System Synthesis (CODES+ISSS '06)*, pp. 211–216, Seoul, Korea, October 2006.

[20] T. Kangas, P. Kukkala, H. Orsila, et al., "UML-based multiprocessor SoC design framework," *ACM Transactions on Embedded Computing Systems*, vol. 5, no. 2, pp. 281–320, 2006.

[21] J. Eker, J. W. Janneck, E. A. Lee, et al., "Taming heterogeneity - the ptolemy approach," *Proceedings of the IEEE*, vol. 91, no. 1, pp. 127–144, 2003.

[22] Cadence, "Incisive-SPW," Cadence Design Systems, 2003, http://www.cadence.com/.

[23] Synopsys, "System Studio—Data Sheet," 2003, http://www.synopsys.com/.

[24] J. Buck and R. Vaidyanathan, "Heterogeneous modeling and simulation of embedded systems in El Greco," in *Proceedings of the 8th International Workshop on Hardware/Software Codesign (CODES '00)*, pp. 142–146, San Diego, Calif, USA, May 2000.

[25] F. Herrera, P. Sánchez, and E. Villar, "Modeling of CSP, KPN and SR systems with SystemC," in *Languages for System Specification: Selected Contributions on UML, SystemC, System Verilog, Mixed-Signal Systems, and Property Specifications from FDL '03*, pp. 133–148, Kluwer Academic, Norwell, Mass, USA, 2004.

[26] H. D. Patel and S. K. Shukla, "Towards a heterogeneous simulation kernel for system-level models: a SystemC kernel for synchronous data flow models," *IEEE Transactions on Computer-Aided Design of Integrated Circuits and Systems*, vol. 24, no. 8, pp. 1261–1271, 2005.

[27] H. D. Patel and S. K. Shukla, "Towards a heterogeneous simulation kernel for system level models: a SystemC kernel for synchronous data flow models," in *Proceedings of the 14th ACM Great Lakes Symposium on VLSI (GLSVLSI '04)*, pp. 248–253, Boston, Mass, USA, April 2004.

[28] H. D. Patel and S. K. Shukla, *SystemC Kernel Extensions for Heterogenous System Modeling*, Kluwer Academic, Norwell, Mass, USA, 2004.

[29] J. Liu, J. Eker, J. W. Janneck, X. Liu, and E. A. Lee, "Actor-oriented control system design: a responsible framework perspective," *IEEE Transactions on Control Systems Technology*, vol. 12, no. 2, pp. 250–262, 2004.

[30] G. Agha, "Abstracting interaction patterns: a programming paradigm for open distribute systems," in *Formal Methods for Open Object-based Distributed Systems*, E. Najm and J.-B. Stefani, Eds., pp. 135–153, Chapman & Hall, London, UK, 1997.

[31] E. A. Lee and D. G. Messerschmitt, "Static scheduling of synchronous data flow programs for digital signal processing," *IEEE Transactions on Computers*, vol. 36, no. 1, pp. 24–35, 1987.

[32] G. Kahn, "The semantics of simple language for parallel programming," in *Proceedings of IFIP Congress*, pp. 471–475, Stockholm, Sweden, August 1974.

[33] JTC 1/SC 29; ISO, "ISO/IEC 14496: Coding of Audio-Visual Objects," Moving Picture Expert Group.

[34] K. Strehl, L. Thiele, M. Gries, D. Ziegenbein, R. Ernst, and J. Teich, "FunState—an internal design representation for

codesign," *IEEE Transactions on Very Large Scale Integration (VLSI) Systems*, vol. 9, no. 4, pp. 524–544, 2001.

[35] E. A. Lee and D. G. Messerschmitt, "Synchronous data flow," *Proceedings of the IEEE*, vol. 75, no. 9, pp. 1235–1245, 1987.

[36] G. Bilsen, M. Engels, R. Lauwereins, and J. Peperstraete, "Cyclo-static dataflow," *IEEE Transactions on Signal Processing*, vol. 44, no. 2, pp. 397–408, 1996.

[37] S. S. Battacharyya, E. A. Lee, and P. K. Murthy, *Software Synthesis from Dataflow Graphs*, Kluwer Academic, Norwell, Mass, USA, 1996.

[38] C.-J. Hsu, S. Ramasubbu, M.-Y. Ko, J. L. Pino, and S. S. Bhattacharvva, "Efficient simulation of critical synchronous dataflow graphs," in *Proceedings of 43rd ACM/IEEE Design Automation Conference (DAC '06)*, pp. 893–898, San Francisco, Calif, USA, July 2006.

[39] Q. Ning and G. R. Gao, "A novel framework of register allocation for software pipelining," in *Conference Record of the 20th Annual ACM SIGPLAN-SIGACT Symposium on Principles of Programming Languages*, pp. 29–42, Charleston, SC, USA, January 1993.

[40] T. M. Parks, J. L. Pino, and E. A. Lee, "A comparison of synchronous and cyclo-static dataflow," in *Proceedings of the 29th Asilomar Conference on Signals, Systems, and Computers*, vol. 1, pp. 204–210, Pacific Grove, Calif, USA, October-November 1995.

[41] V. Pareto, *Cours d' Économie Politique*, vol. 1, F. Rouge & Cie, Lausanne, Switzerland, 1896.

[42] T. Blickle, J. Teich, and L. Thiele, "System-level synthesis using evolutionary algorithms," *Design Automation for Embedded Systems*, vol. 3, no. 1, pp. 23–58, 1998.

[43] IBM, "On-Chip Peripheral Bus—Architecture Specifications," April 2001, Version 2.1.

[44] E. Zitzler, *Evolutionary algorithms for multiobjective optimization: methods and applications*, Ph.D. thesis, Eidgenössische Technische Hochschule Zurich, Zurich, Switzerland, November 1999.

[45] M. Eisenring, L. Thiele, and E. Zitzler, "Conflicting criteria in embedded system design," *IEEE Design and Test of Computers*, vol. 17, no. 2, pp. 51–59, 2000.

[46] K. Deb, *Multi-Objective Optimization Using Evolutionary Algorithms*, John Wiley & Sons, New York, NY, USA, 2001.

[47] T. Schlichter, C. Haubelt, and J. Teich, "Improving EA-based design space exploration by utilizing symbolic feasibility tests," in *Proceedings of Genetic and Evolutionary Computation Conference (GECCO '05)*, H.-G. Beyer and U.-M. O'Reilly, Eds., pp. 1945–1952, Washington, DC, USA, June 2005.

[48] T. Schlichter, M. Lukasiewycz, C. Haubelt, and J. Teich, "Improving system level design space exploration by incorporating SAT-solvers into multi-objective evolutionary algorithms," in *Proceedings of IEEE Computer Society Annual Symposium on Emerging VLSI Technologies and Architectures*, pp. 309–314, Klarlsruhe, Germany, March 2006.

[49] C. Haubelt, T. Schlichter, and J. Teich, "Improving automatic design space exploration by integrating symbolic techniques into multi-objective evolutionary algorithms," *International Journal of Computational Intelligence Research*, vol. 2, no. 3, pp. 239–254, 2006.

[50] M. Streubühr, J. Falk, C. Haubelt, J. Teich, R. Dorsch, and T. Schlipf, "Task-accurate performance modeling in SystemC for real-time multi-processor architectures," in *Proceedings of Design, Automation and Test in Europe (DATE '06)*, vol. 1, pp. 480–481, Munich, Germany, March 2006.

[51] G. C. Buttazzo, *Hard Real-Time Computing Systems*, Kluwer Academic, Norwell, Mass, USA, 2002.

[52] P. Hastono, S. Klaus, and S. A. Huss, "Real-time operating system services for realistic SystemC simulation models of embedded systems," in *Proceedings of the International Forum on Specification & Design Languages (FDL '04)*, pp. 380–391, Lille, France, September 2004.

[53] P. Hastrono, S. Klaus, and S. A. Huss, "An integrated SystemC framework for real-time scheduling. Assessments on system level," in *Proceedings of the 25th IEEE International Real-Time Systems Symposium (RTSS '04)*, pp. 8–11, Lisbon, Portugal, December 2004.

[54] T. Kempf, M. Doerper, R. Leupers, et al., "A modular simulation framework for spatial and temporal task mapping onto multi-processor SoC platforms," in *Proceedings of Design, Automation and Test in Europe (DATE '05)*, vol. 2, pp. 876–881, Munich, Germany, March 2005.

[55] XILINX, *Embedded System Tools Reference Manual—Embedded Development Kit EDK 8.1ia*, October 2005.

[56] S. Klaus, S. A. Huss, and T. Trautmann, "Automatic generation of scheduled SystemC models of embedded systems from extended task graphs," in *System Specification & Design Languages - Best of FDL '02*, E. Villar and J. P. Mermet, Eds., pp. 207–217, Kluwer Academic, Norwell, Mass, USA, 2003.

[57] B. Niemann, F. Mayer, F. Javier, R. Rubio, and M. Speitel, "Refining a high level SystemC model," in *SystemC: Methodologies and Applications*, W. Müller, W. Rosenstiel, and J. Ruf, Eds., pp. 65–95, Kluwer Academic, Norwell, Mass, USA, 2003.

[58] C.-J. Hsu, M.-Y. Ko, and S. S. Bhattacharyya, "Software synthesis from the dataflow interchange format," in *Proceedings of the International Workshop on Software and Compilers for Embedded Systems*, pp. 37–49, Dallas, Tex, USA, September 2005.

[59] P. Lieverse, P. van der Wolf, and E. Deprettere, "A trace transformation technique for communication refinement," in *Proceedings of the 9th International Symposium on Hardware/Software Codesign (CODES '01)*, pp. 134–139, Copenhagen, Denmark, April 2001.

[60] K. Strehl, *Symbolic methods applied to formal verification and synthesis in embedded systems design*, Ph.D. thesis, Swiss Federal Institute of Technology Zurich, Zurich, Switzerland, February 2000.

[61] K. Z. Bukhari, G. K. Kuzmanov, and S. Vassiliadis, "DCT and IDCT implementations on different FPGA technologies," in *Proceedings of the 13th Annual Workshop on Circuits, Systems and Signal Processing (ProRISC '02)*, pp. 232–235, Veldhoven, The Netherlands, November 2002.

[62] C. Loeffer, A. Ligtenberg, and G. S. Moschytz, "Practical fast 1-D DCT algorithms with 11 multiplications," in *Proceedings of IEEE International Conference on Acoustics, Speech, and Signal Processing (ICASSP '89)*, vol. 2, pp. 988–991, Glasgow, UK, May 1989.

[63] J. Liang and T. D. Tran, "Fast multiplierless approximation of the DCT with the lifting scheme," in *Applications of Digital Image Processing XXIII*, vol. 4115 of *Proceedings of SPIE*, pp. 384–395, San Diego, Calif, USA, July 2000.

[64] A. C. Hung and T. H.-Y. Meng, "A comparison of fast inverse discrete cosine transform algorithms," *Multimedia Systems*, vol. 2, no. 5, pp. 204–217, 1994.

Hindawi Publishing Corporation
EURASIP Journal on Embedded Systems
Volume 2007, Article ID 60834, 7 pages
doi:10.1155/2007/60834

Research Article

Priority-Based Heading One Detector in H.264/AVC Decoding

Ke Xu, Chiu-Sing Choy, Cheong-Fat Chan, and Kong-Pang Pun

Department of Electronic Engineering, The Chinese University of Hong Kong, Shatin, Hong Kong

Received 11 July 2006; Accepted 31 January 2007

Recommended by Jarmo Henrik Takala

A novel priority-based heading one detector for Exp-Golomb/CAVLC decoding of H.264/AVC is presented. It exploits the statistical distribution of input encoded codewords and adopts a nonuniform partition decoding scheme for the detector. Compared with a conventional design without power optimization, the power consumption can be reduced by more than 3 times while the performance is maintained and the design hardware cost does not increase. The proposed detector has successfully been verified and implemented in a complete H.264/AVC decoding system.

1. INTRODUCTION

The Moving Picture Experts Group and the Video Coding Experts Group (MPEG and VCEG) have jointly developed a new video coding standard named as H.264/AVC [1]. Compared with previous coding standards like MPEG-2 or H.263, it achieves nearly the same video quality (by means of PSNR and subjective testing) while requiring 60% or less of the bit rate [2]. This substantial improvement comes at a price of extraordinarily huge computational complexity and formidable memory access, which in turn incur greater power consumption.

On the other hand, CMOS technology has now entered the "power-limited scaling regime," where power consumption moves from being one of many design metrics to be number one design metric. The H.264/AVC processing demands much greater power than MPEG-2 or H.263 due to increased complexity. Therefore, its power consumption should be carefully managed to meet power budget, especially for applications on portable devices. Although power dissipation can be substantially reduced through technology scaling, where designers switch to a smaller geometry to implement the same circuit, power reduction through proper design techniques is more flexible and extensive, especially where geometry scaling is not applicable.

H.264/AVC standard defines a hybrid block-based video codec, which is in general similar to early coding standards, but the important changes occur in the details of each functional block with many new coding techniques. One of these techniques occurs in entropy coding, where two methods, Exp-Golomb for syntax elements above the slice layer and CAVLC (context-adaptive variable-length coding) for quantized transform coefficients, are supported in the baseline profile [3]. During the decoding process, all the Exp-Golomb coded syntax elements require the identification of the position of the first appeared "1" inside each codeword. For CAVLC decoding, some parameters like TotalCoeff, level_prefix, and total_zeros tables [1] also need to identify this first "1" before lookup table operation happens.

Conventional detectors usually are not aware of power consumption. One such example is described in [4] which splits the 16-bit input into 4 parts (4-bit vectors), each of which detects whether there is a "1" among the four input bits. Then these results will determine which part should be further tested. Although the method works well, it is not a power-efficient technique since it treats all the 16 input bits with equal importance. The power consumption bears no relationship with the occurrence of any codewords; no matter how likely they will occur.

General low-power design techniques have been developed for many years. Besides these general methods, video decoding presents a unique power optimization opportunity due to temporal, spatial, and statistical redundancies in digital video data. In this paper, we mainly utilize statistical redundancy during video decoding. A data-driven priority-based heading one detector is proposed, which detects the heading "1" in a bitstream that is organized in 16-bit units. The key idea of our proposal is to exploit the statistical characteristics of the heading one position among the various codewords. A nonuniform decoding scheme is designed

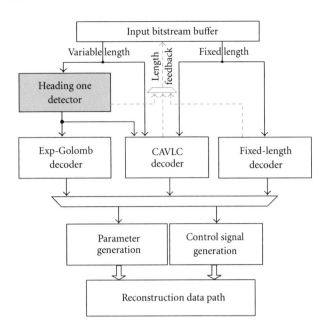

FIGURE 1: Decoder system architecture.

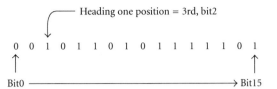

FIGURE 2: Heading one position.

TABLE 1: Exp-Golomb codewords.

Code_num	Codeword
0	1
1	010
2	011
3	00100
4	00101
5	00110
6	00111
7	0001000
8	0001001
⋯	[M zeros][1][INFO]

accordingly. By selectively disabling some subblocks, the detector consumes much less power without noticeable performance degradation and even smaller design area.

2. BACKGROUND

In this section, we firstly give a brief introduction of the whole decoder architecture. Then we discuss the structure of Exp-Golomb code and CAVLC code which requires heading one detection. At last, we evaluate the related research works in literature.

2.1. H.264/AVC decoding

A simplified system architecture of the whole decoder is illustrated in Figure 1. According to input codeword type, the heading one detector is invoked when current codeword is Exp-Golomb coded or a certain part of CAVLC codewords. Based on the output of the heading one detector, Exp-Golomb codes are mapped from bitstream form to signed, unsigned, or truncated syntax element values, while CAVLC codes are indexed for several lookup Tables (LUT). There is a length feedback signal from the heading one detector, CAVLC decoder, and fixed-length decoder to the input bitstream buffer. The signal indicates how many bits are consumed for decoding current codeword. According to decoded codewords, related parameters and control signals are generated to orchestrate the following reconstruction data path.

2.2. Heading one detection

Figure 2 depicts a normal input to the heading one detector, where the detector needs to search among the 16 bits to find

the first appeared "1." Here we assume the input bitstream is encoded from left to right. This example indicates that the heading one position lies at third place (bit2). Although there are several "1's" at some other positions like bit4, bit5, and so forth, they are not heading ones.

Exp-Golomb codes

Exponential Golomb codes (see [5]) are variable-length codes with simple and regular structure as depicted in Table 1. One does not need to store the conversion table for the purpose of decoding, since the correspondence between symbols and codes is mathematically defined. The leading M zeros, as well as the middle "1," are treated as "prefix" of the codeword, while INFO, which is equal in length to the M zeros, is called "suffix" [6]. In Table 1, the first code_num "0" does not contain any leading zero or trailing INFO. Code_nums "1" and "2" have a single-bit leading zero and corresponding single-bit INFO field, code_nums 3~6 have a two-bit leading zeros and INFO field, and so on. Theoretically the codeword table can be infinitely extended according to the coding rule described. The length of each Exp-Golomb codeword is $(2M + 1)$ bits long and each codeword can be inferred by the following equation [6]:

$$M = \text{floor}(\log_2[\text{code\_num} + 1]),$$
$$\text{INFO} = \text{code\_num} + 1 - 2^M, \tag{1}$$

where floor (x) is a function finding the largest integer which is less than or equal to x.

In H.264/AVC standard, there are three types of Exp-Golomb coding: unsigned, signed, and truncated. They all follow the same coding rule and are only different in whether an additional "code_num to syntax_value" mapping is needed.

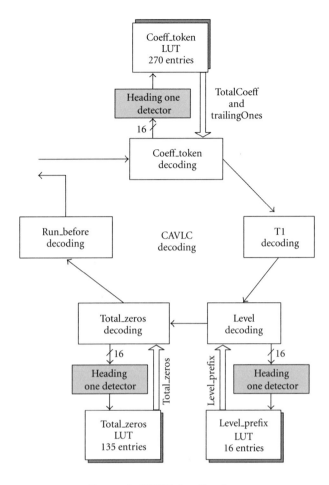

FIGURE 3: CAVLC decoding flow.

TABLE 2: Codeword table for level_prefix.

level_prefix	Bit string
0	1
1	01
2	001
3	0001
4	0000 1
5	0000 01
6	0000 001
7	0000 0001
. . .	. . .

CAVLC

A more efficient algorithm for transmitting the quantized transform coefficients is proposed in [3]. In this method, VLC tables for various syntax elements are selected depending on already transmitted syntax elements. To decode the indexes for some of these VLC tables, a heading one detector is indispensable. The CAVLC decoding step is briefly described in Figure 3.

The CAVLC decoding can be partitioned into five steps and three of them require heading one detection.

TABLE 3: Total_zeros table for 4×4 blocks with TotalCoeff (coeff_token) 1 to 3.

Total_zeros	TotalCoeff (coeff_token)		
	1	2	3
0	1	111	0101
1	011	110	111
2	010	101	110
3	0011	100	101
4	0010	011	0100
5	0001 1	0101	0011
6	0001 0	0100	100
7	0000 11	0011	011
. . .	. . .	. . .	. . .

Table 2 shows one VLC table [1] in CAVLC codes which maps input bit stream to "level_prefix." The value of level_prefix is directly determined by the position of the first appeared "1." Table 3 shows another VLC example where finding the heading one position is sufficient for the whole syntax element to be extracted.

Since most of the syntax elements are coded either as Exp-Golomb codes or CAVLC codes, heading one detector is used extensively in H.264/AVC decoding.

2.3. Related works

Although there are some designs in literature dealing with Exp-Golomb or CAVLC decoding [4, 7–9], few of them mentioned how heading one detection was realized. The only reference design is found in [4]. It proposed a detector that evenly splits the input into four subwords. From each subword, the presence of "1" is detected. Then these results will determine which subword should be further tested, as shown in Figure 4. Priority encoder0's output indicates the position of "1" in the subword, while priority encoder1's output indicates which subword has the heading one. In fact, this is a two-level encoder and cannot run in parallel. Encoder1 selects a subword based on priority where part [3 : 0] has the highest priority and part [15 : 12] has the lowest priority. According to encoder1's indication, encoder0 chooses one correct subword among the four and encodes the heading "1" in the chosen subword as the final heading one position. No matter where the heading one is, four subword decoders and two priority encoders are active all the time.

3. PROPOSED ARCHITECTURE

In this section, we firstly explore the heading one statistics in entropy coding. Based on the observation, a priority-based heading one detector is then proposed.

3.1. Characteristic of entropy coding

As aforementioned, design in [4] proposed a "first 1 detector" based on a uniform input bit-vector partition. That is an effective scheme but no power optimization was considered.

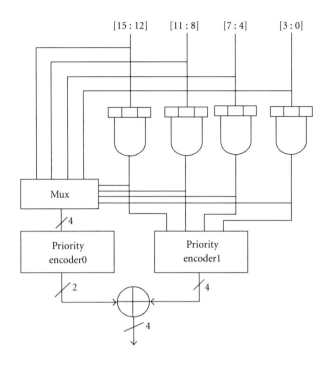

FIGURE 4: Evenly partitioned detector in [4].

TABLE 4: Statistic result of heading one position (nearly 0 means that percentage is less than 0.01%).

Position	Whole input bitstream	Intracoded frame	Intercoded frame
0	55.36%	51.37%	56.61%
1	24.15%	24.36%	24.08%
2	11.16%	10.70%	11.31%
3	5.49%	6.21%	5.26%
4	2.17%	3.69%	1.72%
5	0.88%	1.6%	0.65%
6	0.41%	0.91%	0.25%
7	0.16%	0.4%	0.08%
8	0.08%	0.25%	0.03%
9	0.06%	0.24%	0.01%
10	0.04%	0.15%	Nearly 0
11	0.01%	0.06%	Nearly 0
12	Nearly 0	0.04%	Nearly 0
13	Nearly 0	0.03%	Nearly 0
14	Nearly 0	Nearly 0	Nearly 0
15	Nearly 0	Nearly 0	Nearly 0
Average	0.81	1.12	0.74

Since both Exp-Golomb and CAVLC codings are entropy coding methods, they have the same important characteristic like all other entropy coding schemes: shorter codewords are assigned to symbols that occur with higher probability, whereas longer codewords are assigned to symbols with less frequent occurrences. In an H.264/AVC bitstream, the longest code is 16 bits including the heading "1." However, the average length of such kind of codes is not $(16 + 1)/2 = 8.5$, but much smaller.

3.2. SystemC modeling

In order to study the entire bitstream parsing process where entropy decoding is included, we developed a high-level systemC model, emulating the control and communication of real video decoding. Its output is compared with JM9.4 software [10] to verify correct function. The systemC model has internal counters to count the total number of Exp-Golomb codes and CAVLC codes which require heading one detection. It also has individual counters for the number of these codes under different heading one positions. Five popular test videos, named as container, foreman, akiyo, news and carphone, with QCIF 300 frame sequences at 30 fps are used. They are encoded by JM software with quantization parameters set to 22, 25, 28, 32, and 36, respectively. The statistical profile of heading one's positions was hence obtained from simulation with these input bitstreams.

The average codeword lengths are found as in Table 4 (note that if a "1" is in the first bit, this corresponds to position = 0 and so on). The intraframe and interframe have slightly different heading one statistical position percentage since usually the intraframe has more residual information and needs more CAVLC decoding effort. For example, inside interframe, positions equal to or above 10 begin to have nearly zero (less than 0.01%) codes distribution, whereas for intra frame, this boundary is pushed to a high position which indicates that only positions 14 and 15 have nearly zero codes distribution. However, both intra- and interframes-share the same tendency that the higher the position is, the less opportunity that a heading one is found.

Be aware that the statistical positions stated in Table 2 are not a simple average of the values in the intra- and the interframe columns. This is because intra- and interframes have different total numbers of Exp-Golomb/CAVLC codes in different test video sequences. For example, in akiyo video sequence of 300 frames, 24% of codes need heading one detection are extracted from intraframes and 76% are extracted from interframes, while in foreman video sequence, 30% of these codes are extracted from intraframes and 70% are from inter frames. In addition, distributions of heading one positions (position = $0, 1, 2, \ldots$) in a single video sequence vary from one bitstream to another. These nonuniform codewords distributions lead to the nonlinear relationship of total average positions for intra- and interframes. In addition, positions of interframes tend to have a larger weight than those of intraframes in all the video sequences tested, for there are more intercoded frames than intra-coded ones.

According to Table 4, the heading one in a codeword is located on average in a position indicated in Figure 5. We conclude that the average heading one position for the whole sequence/intraframe/interframe is 0.81/1.12/0.74, respectively, which are much smaller than the simple average of 8.5. Of course, positions naturally are whole numbers, fractional values are the artifacts of averaging.

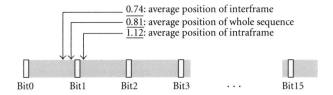

FIGURE 5: Average heading one position.

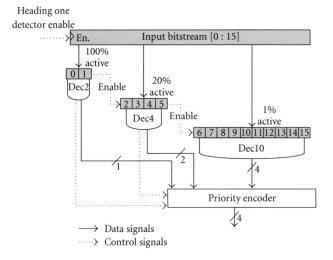

FIGURE 6: Proposed heading one detector.

3.3. Proposed architecture

From the above analysis, we conclude that the position of heading one most likely lies around the second input bit (position = 1). The first two positions (position0 + position1) account for almost 80% of all cases and the first six positions (position0 + ⋯ + position5) account for nearly 99%. Thus we propose a priority-based nonuniform partition heading one detector where the input 16 bits are divided into 3 unequal subdetectors and each subdetector can be selectively enabled and disabled. Figure 6 shows the proposed scheme.

In our design, input bitstream from bitstream buffer is controlled by an "enable" signal. If current codeword needs heading one detection, the whole 16 bits are enabled and passed to the heading one detector, else the detector is disabled to reduce unnecessary switching. The entire detector is partitioned into three parts, each of which handles a different chunk of input bits with varying priority. Dec2, which has the highest priority, processes the first two input bits and is active all the time to detect whether there is a "1" and its corresponding likely position (only first bit position or second bit position here). If a "1" is found in dec2, which signals a successful identification of the heading "1" in a codeword, the position information is passed to the final priority encoder to generate a heading one position. At the same time, the lower-priority dec4 and the lowest-priority dec10 are disabled to save power. Conversely, there is a 20% possibility that dec2 will fail to find a "1" and dec4 will be enabled there-

after. More rarely, both dec2 and dec4 cannot find a heading one and dec10 will then be active but having only 1% possibility. The outputs of dec2, dec4, and dec10 are selectively encoded as heading one position of whole 16-bit input by a priority encoder.

Design in [4] divides 16-bits input evenly into 4 identical subwords. Each subword decoder detects whether there is a "1" inside and their outputs are then sent to two priority encoders. No matter whether the "1" found in each subword is a heading "1," all four subword decoders, as well as the priority encoders, are active all the time. However, if the first decoder which looks at bits [3 : 0] finds a "1," no matter what the outcome of the other three decoders is, one can conclude that the first "1" is in bits [3 : 0]. The work done by the other three decoders is of no consequence and only a waste of power.

4. DESIGN ANALYSIS

In this section, we mainly discuss and compare the power consumption of [4] and the proposed design. We also discuss the speed and area overheads.

4.1. Theoretical analysis

Strictly speaking, power consumption constitutes of dynamic power and static power. Since the target process is a relatively standard CMOS 130 nm technology and the circuit is small enough, the static power only contributes a very small portion of the whole power consumption. Therefore, we can assume that the detector's entire power is proportional to dynamic power to facilitate our calculation. Average power dissipation for decoding each heading one position can be modeled as suggested in [11]

$$E_{\text{avg}} = \sum_{i=1}^{N} P_i E_i, \qquad (2)$$

where P_i is the probability that heading position = i will occur, E_i is the energy required to detect such a position, and N is the total number of possible positions where $N = 16$ for H.264/AVC codes.

Since dynamic power consumption is almost linear to the complexity of these decoding units, without loss of generality, one can assume the power consumed by dec2 is 2 units, dec4 is 4 units, and dec10 is 10 units. In [4], all the four decoders are identical and consume 4 units of power all the time.

Estimated power consumption for the detector in [4] is:

$$E_{\text{avg}} = \sum_{i=1}^{4} P_i E_i = 4 \times 100\% \times 4 \text{ units} = 16 \text{ units}. \qquad (3)$$

In our scheme, three decoders are active sequentially and their activation rate is proportional to the heading one distribution shown in Table 4.

TABLE 5: Layout power analysis.

| Frame type | Power consumption at 20 MHz real-time QCIF/30 fps | | |
	Implementation of [3]	Our proposal	Power reduction
Intra frame	13.45 μW	3.99 μW	3.38 times
Inter frame	2.35 μW	0.733 μW	3.21 times

TABLE 6: Physical implementation.

Technology	UMC 130 nm
Metal layer	6 metals, 2 thick
Supply voltage	1.08 v
Max. frequency	200 MHz

Estimated power consumption for our scheme is

$$
\begin{aligned}
E_{avg} &= \sum_{i=1}^{3} P_i E_i \\
&= 100\% \times 2 \text{ units} + 20\% \times 4 \text{ units} + 1\% \times 10 \text{ units} \\
&= 2.9 \text{ units.}
\end{aligned}
\tag{4}
$$

The percentages in the above equation reflect the activity rate of each submodule dec2 (for position 0~1), dec4 (for position 2~5), and dec10 (for other positions), respectively. The overhead-like power consumed by muxes is negligible. The relative power saving for our scheme is about 5.5 times while the throughput is nearly the same.

4.2. Implementation analysis

Since there is no power consumption figures reported in [4], to have a fair comparison, we built a "heading 1 detector" according to [4] with the same process technology used for our scheme. Both of the detectors are integrated into H.264/AVC decoding system, where there is a switch to control which one is currently active. The decoding system is simulated by ModelSim. The Verilog RTL codes are then synthesized by design compiler and are placed and routed by Astro. Parasitic information is extracted by Star-RCXT and postsimulation is processed in VCS. Based on the layout database and individual activity rate obtained from post-sim, postlayout power analysis results can be obtained from PrimePower, shown in Table 5. The key implementation parameters of our scheme are listed in Table 6. Considering that the heading one detector has the highest switching activity in entropy decoding, the power reduction contributable by such a detector is substantial.

According to Tables 5 and 6, one can conclude that our design not only consumes less power, but is capable of performing real-time decoding. The circuit size is even a little bit smaller than the design in [4]. Although a larger dec10 is introduced, two priority encoders found in [4] are reduced to one which leads to slight area reduction. The only penalty is a small throughput degradation if the heading one happened to be at a higher position like 6, 7, and so forth, because dec2, dec4, and dec10 will need to be triggered in sequence to obtain the final result. Even at this extreme case, the proposed design can achieve a maximum frequency of 200 MHz, which is substantially faster than other building blocks in the whole H.264/AVC decoding system.

The advantage of our design is drawn from exploiting the high probability of "heading one" lying in the first few bits of a codeword. By using a nonuniform decoding structure, a lot of power is saved because one does not need to search all bits. The same technique can also be applied to other entropy decodings such as that in MPEG-2. Although the codeword structure is not identical as in H.264/AVC, short codewords inherently occur more frequently. Proposed technique can then be employed according to the specific statistical profile found from high-level modeling.

5. CONCLUSION

A priority-based, data-driven power-efficient heading one detector has been proposed. The opportunity to reduce power is identified at architectural level through systemC modeling. Appropriate circuit implementation is then chosen. It exploits the statistical codeword distribution of an entropy-coded bitstream, and a novel power-saving decoding scheme is subsequently devised. Compared with conventional detectors, the proposed design achieves more than 3 times power reduction while maintaining area and speed performance. It does not utilize any special techniques such as clock gating or voltage scaling, and thus makes it readily employable in other circumstances when different technologies may be used. Since power consumption in ICs is a critical issue in recent years, this paper suggests an effective method to reduce power by exploiting statistical characteristics.

ACKNOWLEDGMENT

The work reported is supported by a Hong Kong SAR Government Research Direct Grant no. 2050322.

REFERENCES

[1] J. V. Team, "Advanced video coding for generic audiovisual services," *ITU-T Recommendation H.264 and ISO/IEC 14496-10 AVC*, May 2003.

[2] T. Wiegand, H. Schwarz, A. Joch, F. Kossentini, and G. J. Sullivan, "Rate-constrained coder control and comparison of video coding standards," *IEEE Transactions on Circuits and Systems for Video Technology*, vol. 13, no. 7, pp. 688–703, 2003.

[3] T. Wiegand, G. J. Sullivan, G. Bjntegaard, and A. Luthra, "Overview of the H.264/AVC video coding standard," *IEEE Transactions on Circuits and Systems for Video Technology*, vol. 13, no. 7, pp. 560–576, 2003.

[4] W. Di, G. Wen, H. Mingzeng, and J. Zhenzhou, "An Exp-Golomb encoder and decoder architecture for JVT/AVS," in *Proceedings of the 5th International Conference on ASIC*, vol. 2, pp. 910–913, Beijing, China, October 2003.

[5] S. W. Golomb, "Run-length encoding," *IEEE Transactions on Information Theory*, vol. 12, no. 3, pp. 399–401, 1966.

[6] I. E. G. Richardson, *H.264 and MPEG-4 Video Compression*, John Willey & Sons, New York, NY, USA, 2003.

[7] Joint Video Team (JVT) reference software JM9.4, http://iphome.hhi.de/suehring/tml/download/.

[8] T.-C. Wang, H.-C. Fang, W.-M. Chao, H.-H. Chen, and L.-G. Chen, "An UVLC encoder architecture for H.26L," in *Proceedings of IEEE International Symposium on Circuits and Systems (ISCAS '02)*, vol. 2, pp. 308–311, Phoenix, Ariz, USA, May 2002.

[9] S. H. Cho, T. Xanthopoulos, and A. P. Chandrakasan, "A low power variable length decoder for MPEG-2 based on nonuniform fine-grain table partitioning," *IEEE Transactions on VLSI Systems*, vol. 7, no. 2, pp. 249–257, 1999.

[10] I. Amer, W. Badawy, and G. Jullien, "Towards MPEG-4 part 10 system on chip: a VLSI prototype for context-based adaptive variable length coding (CAVLC)," in *Proceedings of IEEE Workshop on Signal Processing Systems (SIPS '04)*, pp. 275–279, Austin, Tex, USA, October 2004.

[11] H.-Y. Lin, Y.-H. Lu, B.-D. Liu, and J.-F. Yang, "Low power design of H.264 CAVLC decoder," in *Proceedings of IEEE International Symposium on Circuits and Systems (ISCAS '06)*, pp. 2689–2692, Island of Kos, Greece, May 2006.